KU-178-241

2016 SQA Past Papers With Answers

National 5 GEOGRAPHY

2014, 2015 & 2016 Exams

This book contains the official SQA 2014, 2015 and 2016 Exams for National 5 Geography, with associated SQA-approved answers modified from the official marking instructions that accompany the paper.

In addition the book contains study skills advice. This advice has been specially commissioned by Hodder Gibson, and has been written by experienced senior teachers and examiners in line with the new National 5 syllabus and assessment outlines. This is not SQA material but has been devised to provide further guidance for National 5 examinations.

Hodder Gibson is grateful to the copyright holders, as credited on the final page of the Answer Section, for permission to use their material. Every effort has been made to trace the copyright holders and to obtain their permission for the use of copyright material. Hodder Gibson will be happy to receive information allowing us to rectify any error or omission in future editions.

Hachette UK's policy is to use papers that are natural, renewable and recyclable products and made from wood grown in sustainable forests. The logging and manufacturing processes are expected to conform to the environmental regulations of the country of origin.

Orders: please contact Bookpoint Ltd, 130 Park Drive, Milton Park, Abingdon, Oxon OX14 4SE. Telephone: (44) 01235 827720. Fax: (44) 01235 400454. Lines are open 9.00–5.00, Monday to Saturday, with a 24-hour message answering service. Visit our website at www.hoddereducation.co.uk. Hodder Gibson can be contacted direct on: Tel: 0141 333 4650; Fax: 0141 404 8188; email: hoddergibson@hodder.co.uk

This collection first published in 2016 by
Hodder Gibson, an imprint of Hodder Education,
An Hachette UK Company
211 St Vincent Street
Glasgow G2 5QY

National 5 2014, 2015 and 2016 Exam Papers and Answers © Scottish Qualifications Authority. Study Skills section © Hodder Gibson. All rights reserved. Apart from any use permitted under UK copyright law, no part of this publication may be reproduced or transmitted in any form or by any means, electronic or mechanical, including photocopying and recording, or held within any information storage and retrieval system, without permission in writing from the publisher or under licence from the Copyright Licensing Agency Limited. Further details of such licences (for reprographic reproduction) may be obtained from the Copyright Licensing Agency Limited, Saffron House, 6–10 Kirby Street, London EC1N 8TS.

Typeset by Aptara, Inc.

Printed in the UK

A catalogue record for this title is available from the British Library

ISBN: 978-1-4718-9110-6

3 2 1

2017 2016

Introduction

Study Skills – what you need to know to pass exams!

Pause for thought

Many students might skip quickly through a page like this. After all, we all know how to revise. Do you really though?

Think about this:

"IF YOU ALWAYS DO WHAT YOU ALWAYS DO, YOU WILL ALWAYS GET WHAT YOU HAVE ALWAYS GOT."

Do you like the grades you get? Do you want to do better? If you get full marks in your assessment, then that's great! Change nothing! This section is just to help you get that little bit better than you already are.

There are two main parts to the advice on offer here. The first part highlights fairly obvious things but which are also very important. The second part makes suggestions about revision that you might not have thought about but which WILL help you.

Part 1

DOH! It's so obvious but ...

Start revising in good time

Don't leave it until the last minute – this will make you panic.

Make a revision timetable that sets out work time AND play time.

Sleep and eat!

Obvious really, and very helpful. Avoid arguments or stressful things too – even games that wind you up. You need to be fit, awake and focused!

Know your place!

Make sure you know exactly **WHEN and WHERE** your exams are.

Know your enemy!

Make sure you know what to expect in the exam.

How is the paper structured?

How much time is there for each question?

What types of question are involved?

Which topics seem to come up time and time again?

Which topics are your strongest and which are your weakest?

Are all topics compulsory or are there choices?

Learn by DOING!

There is no substitute for past papers and practice papers – they are simply essential! Tackling this collection of papers and answers is exactly the right thing to be doing as your exams approach.

Part 2

People learn in different ways. Some like low light, some bright. Some like early morning, some like evening / night. Some prefer warm, some prefer cold. But everyone uses their BRAIN and the brain works when it is active. Passive learning – sitting gazing at notes – is the most INEFFICIENT way to learn anything. Below you will find tips and ideas for making your revision more effective and maybe even more enjoyable. What follows gets your brain active, and active learning works!

Activity 1 – Stop and review

Step 1

When you have done no more than 5 minutes of revision reading STOP!

Step 2

Write a heading in your own words which sums up the topic you have been revising.

Step 3

Write a summary of what you have revised in no more than two sentences. Don't fool yourself by saying, "I know it, but I cannot put it into words". That just means you don't know it well enough. If you cannot write your summary, revise that section again, knowing that you must write a summary at the end of it. Many of you will have notebooks full of blue/black ink writing. Many of the pages will not be especially attractive or memorable so try to liven them up a bit with colour as you are reviewing and rewriting. **This is a great memory aid, and memory is the most important thing.**

Activity 2 – Use technology!

Why should everything be written down? Have you thought about "mental" maps, diagrams, cartoons and colour to help you learn? And rather than write down notes, why not record your revision material?

What about having a text message revision session with friends? Keep in touch with them to find out how and what they are revising and share ideas and questions.

Why not make a video diary where you tell the camera what you are doing, what you think you have learned and what you still have to do? No one has to see or hear it, but the process of having to organise your thoughts in a formal way to explain something is a very important learning practice.

Be sure to make use of electronic files. You could begin to summarise your class notes. Your typing might be slow, but it will get faster and the typed notes will be easier to read than the scribbles in your class notes. Try to add different fonts and colours to make your work stand out. You can easily Google relevant pictures, cartoons and diagrams which you can copy and paste to make your work more attractive and **MEMORABLE**.

Activity 3 – This is it. Do this and you will know lots!

Step 1

In this task you must be very honest with yourself! Find the SQA syllabus for your subject (www.sqa.org.uk). Look at how it is broken down into main topics called MANDATORY knowledge. That means stuff you MUST know.

Step 2

BEFORE you do ANY revision on this topic, write a list of everything that you already know about the subject. It might be quite a long list but you only need to write it once. It shows you all the information that is already in your long-term memory so you know what parts you do not need to revise!

Step 3

Pick a chapter or section from your book or revision notes. Choose a fairly large section or a whole chapter to get the most out of this activity.

With a buddy, use Skype, Facetime, Twitter or any other communication you have, to play the game "If this is the answer, what is the question?". For example, if you are revising Geography and the answer you provide is "meander", your buddy would have to make up a question like "What is the word that describes a feature of a river where it flows slowly and bends often from side to side?".

Make up 10 "answers" based on the content of the chapter or section you are using. Give this to your buddy to solve while you solve theirs.

Step 4

Construct a wordsearch of at least 10 × 10 squares. You can make it as big as you like but keep it realistic. Work together with a group of friends. Many apps allow you to make wordsearch puzzles online. The words and phrases can go in any direction and phrases can be split. Your puzzle must only contain facts linked to the topic you are revising. Your task is to find 10 bits of information to hide in your puzzle, but you must not repeat information that you used in Step 3. DO NOT show where the words are. Fill up empty squares with random letters. Remember to keep a note of where your answers are hidden but do not show your friends. When you have a complete puzzle, exchange it with a friend to solve each other's puzzle.

Step 5

Now make up 10 questions (not "answers" this time) based on the same chapter used in the previous two tasks. Again, you must find NEW information that you have not yet used. Now it's getting hard to find that new information! Again, give your questions to a friend to answer.

Step 6

As you have been doing the puzzles, your brain has been actively searching for new information. Now write a NEW LIST that contains only the new information you have discovered when doing the puzzles. Your new list is the one to look at repeatedly for short bursts over the next few days. Try to remember more and more of it without looking at it. After a few days, you should be able to add words from your second list to your first list as you increase the information in your long-term memory.

FINALLY! Be inspired...

Make a list of different revision ideas and beside each one write **THINGS I HAVE** tried, **THINGS I WILL** try and **THINGS I MIGHT** try. Don't be scared of trying something new.

And remember – "FAIL TO PREPARE AND PREPARE TO FAIL!"

National 5 Geography

The exam

The course assessment will consist of two components: a question paper and an assignment.

The question paper

The purpose of this question paper is to assess your application of skills, and breadth of knowledge and understanding across the three units of the course.

This question paper will give you an opportunity to demonstrate the following higher-order cognitive skills and knowledge and understanding from the mandatory content of the course:

- using a limited range of mapping skills;
- using a limited range of numerical and graphical information;
- giving detailed descriptions and explanations with some analysis.

The question paper will have 60 marks (75% of the total mark) distributed across three sections.

Section 1: Physical Environments

This section will have 20 marks and will be made up of limited/extended-response questions. These require you to draw on your knowledge and understanding, and to apply the skills you have acquired during the course.

Section 2: Human Environments

This section will have 20 marks and will be made up of limited/extended-response questions. These require you to draw on your knowledge and understanding, and to apply the skills you have acquired during the course.

These questions will draw on the knowledge and understanding and skills described in the "further mandatory information on course coverage" section.

Section 3: Global Issues

This section will have 20 marks and will be made up of limited/extended-response questions. These require you to draw on your knowledge and understanding, and to apply the skills you have acquired during the course.

These questions will draw on the knowledge and understanding and skills described in the "further mandatory information on course coverage" section. In this section you will be required to attempt two questions from six. The choice of topics are: Climate Change; Impact of Human Activity on the Natural Environment; Environmental Hazards; Trade and Globalisation; Tourism; and Health.

What you will be tested on

For marks to be given, points must relate to the question asked.

There are six types of question used in this paper:

A. Describe
B. Explain
C. Give reasons
D. Match
E. Give map evidence
F. Give advantages and/or disadvantages

*Questions which ask candidates to **describe**:*

You must make a number of relevant, factual points. These should be key points taken from a given source, for example a map, diagram or table.

*Questions which ask candidates to **explain** or **give reasons**:*

You should make a number of points giving clear reasons for a given situation. The command word "explain" will be used when you are asked to demonstrate knowledge and understanding. Sometimes the command words "give reasons" may be used as an alternative to "explain".

*Questions which ask candidates to **match**:*

You are asked to match two sets of variables, for example to match features to a correct grid reference.

*Questions which ask candidates to **give map evidence**:*

You should look for evidence on the map and make clear statements to support your answer.

*Questions which ask candidates to **give advantages and/or disadvantages**:*

You should select relevant advantages or disadvantages of a proposed development, for example the location of a new shopping centre, and demonstrate your understanding of the significance of the proposal.

Some tips for revising

- To be best prepared for the examination, organise your notes into sections. Try to work out a schedule for studying with a programme which includes the sections of the syllabus you intend to study.
- Organise your notes into checklists and revision cards.
- Try to avoid leaving your studying to a day or two before the exam. Also try to avoid cramming your studies into the night before the examination, and especially avoid staying up late to study.
- One useful technique when revising is to use summary note cards on individual topics.

- Make use of past paper questions to test your knowledge or enquiry skills. Go over your answers and give yourself a mark for every correct point you make when comparing your answer with your notes.
- If you work with a classmate, try to mark each other's practice answers.
- Practise your diagram-drawing skills and your writing skills. Ensure that your answers are clearly worded. Try to develop the points that you make in your answers.

Some tips for the exam

- Do not write lists, even if you are running out of time. You will lose marks. If the question asks for an opinion based on a choice, for example on the suitability of a particular site or area for a development, do not be afraid to refer to negative points such as why the alternatives are not as good. You will get credit for this.
- Make sure you have a copy of the examination timetable and have planned a schedule for studying.
- Arrive at the examination in plenty of time with the appropriate equipment – pen, pencil, rubber and ruler.
- Carefully read the instructions on the paper and at the beginning of each part of the question.
- Answer all of the compulsory questions in each paper you sit.
- Use the number of marks as a guide to the length of your answer.
- Try to include examples in your answer wherever possible. If asked for diagrams, draw clear, labelled diagrams.
- Read the question instructions very carefully. If the question asks you to "describe", make sure that this is what you do.
- If you are asked to "explain", you must use phrases such as "due to", "this happens because" and "this is a result of". If you describe rather than explain, you will lose most of the marks for that question.
- If you finish early, do not leave the exam. Use the remaining time to check your answers and go over any questions which you have partially answered, especially Ordnance Survey map questions.
- Practise drawing diagrams which may be included in your answers, for example corries or pyramidal peaks.
- Make sure that you have read the instructions on the question carefully and that you have avoided needless errors. For example, answering the wrong sections or failing to explain when asked to, or perhaps omitting to refer to a named area or case study.
- One technique which you might find helpful, especially when answering long questions worth 10 or more marks, is to "brainstorm" possible points for your answer. You can write these down in a list at the start of your answer. As you go through your answer, you can double-check with your list to ensure that you have put as much into your answer as you can. This stops you from coming out of the exam and being annoyed that you forgot to mention an important point.

Common errors

Markers of the external examination often remark on errors which occur frequently in candidates' answers. These include the following:

Lack of sufficient detail

- Many candidates fail to provide sufficient detail in answers, often by omitting reference to specific examples, or not elaborating or developing points made in their answer. As noted above, a good guide to the amount of detail required is the number of marks given for the question. If, for example, the total marks offered is 6, then you should make at least six valid points.

Listing

- If you write a simple list of points rather than fuller statements in your answer, you will automatically lose marks. For example, in a 4 mark question, you will obtain only 1 mark for a list.
- The same rule applies to a simple list of bullet points. However, if you couple bullet points with some detailed explanation, you could achieve full marks.

Irrelevant answers

- You must read the question instructions carefully so as to avoid giving answers which are irrelevant to the question. For example, if you are asked to "explain" and you simply "describe", you will lose marks. If you are asked for a named example and you do not provide one, you will forfeit marks.

Repetition

- You should be careful not to repeat points already made in your answer. These will not gain any further marks. You may feel that you have written a long answer, but it may contain the same basic information repeated again and again. Unfortunately, these repeated statements will be ignored by the marker.

Good luck!

Remember that the rewards for passing National 5 Geography are well worth it! Your pass will help you to get the future you want for yourself. In the exam, be confident in your own ability. If you're not sure how to answer a question, trust your instincts and just give it a go anyway. Keep calm and don't panic! GOOD LUCK!

NATIONAL 5

2014

X733/75/01 **Geography**

THURSDAY, 29 MAY

9:00 AM – 10:30 AM

Total marks — 60

SECTION 1 — PHYSICAL ENVIRONMENTS — 20 marks

Attempt EITHER question 1 **OR** question 2. **ALSO** attempt questions 3 and 4.

SECTION 2 — HUMAN ENVIRONMENTS — 20 marks

Attempt questions 5, 6, 7 and 8

SECTION 3 — GLOBAL ISSUES — 20 marks

Attempt any TWO of the following

Question 9 — Climate Change
Question 10 — Impact of Human Activity on the Natural Environment
Question 11 — Environmental Hazards
Question 12 — Trade and Globalisation
Question 13 — Tourism
Question 14 — Health

Write your answers clearly in the answer booklet provided. In the answer booklet you must clearly identify the question number you are attempting.

Use **blue** or **black** ink.

Credit will always be given for appropriately labelled sketch maps and diagrams.

Before leaving the examination room you must give your answer booklet to the Invigilator; if you do not, you may lose all the marks for this paper.

©

Extract No 2072/15 & 16

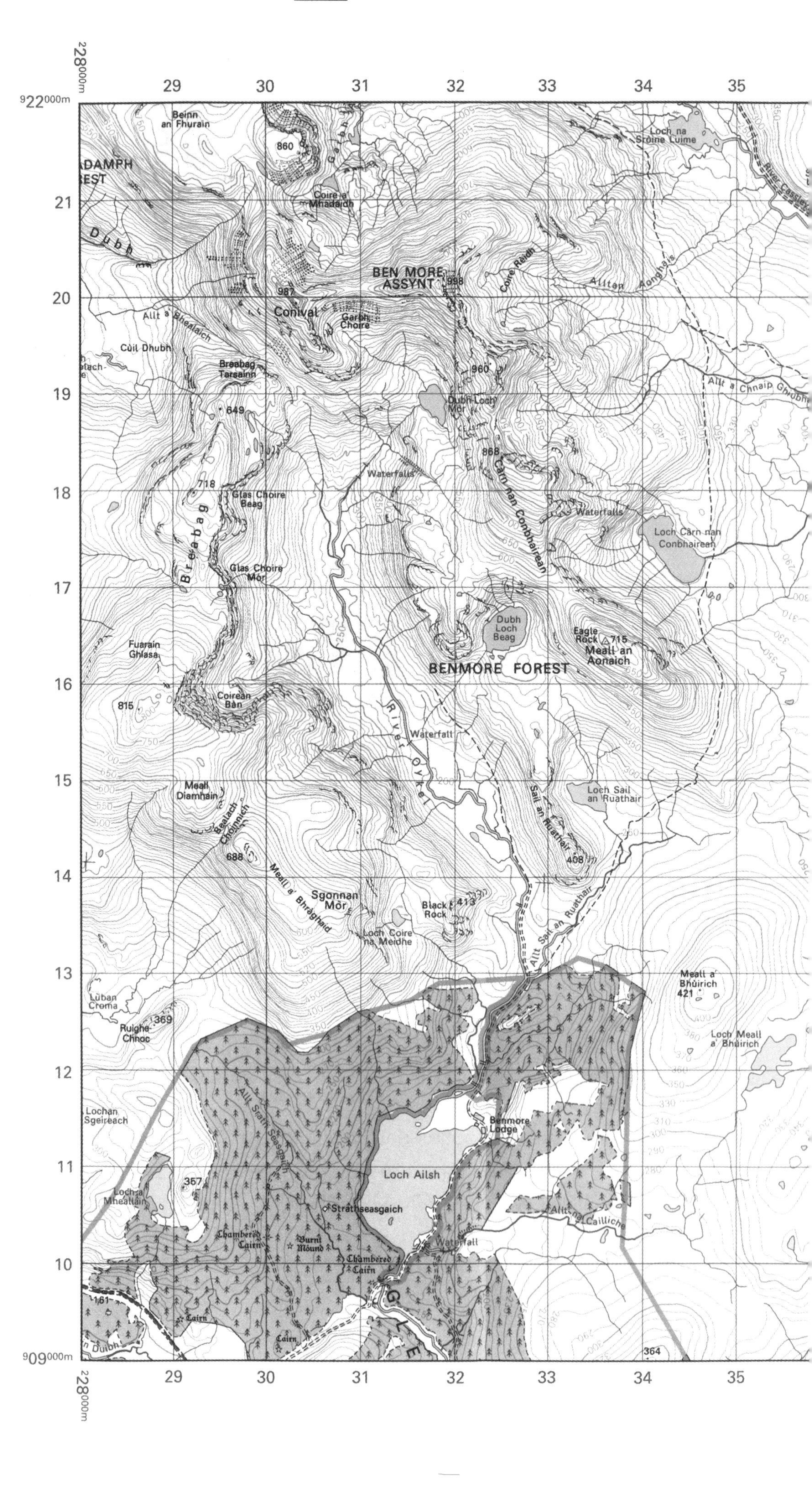

Four colours should appear above; if not then please return to the invigilator.

1:50 000 Scale
Landranger Series

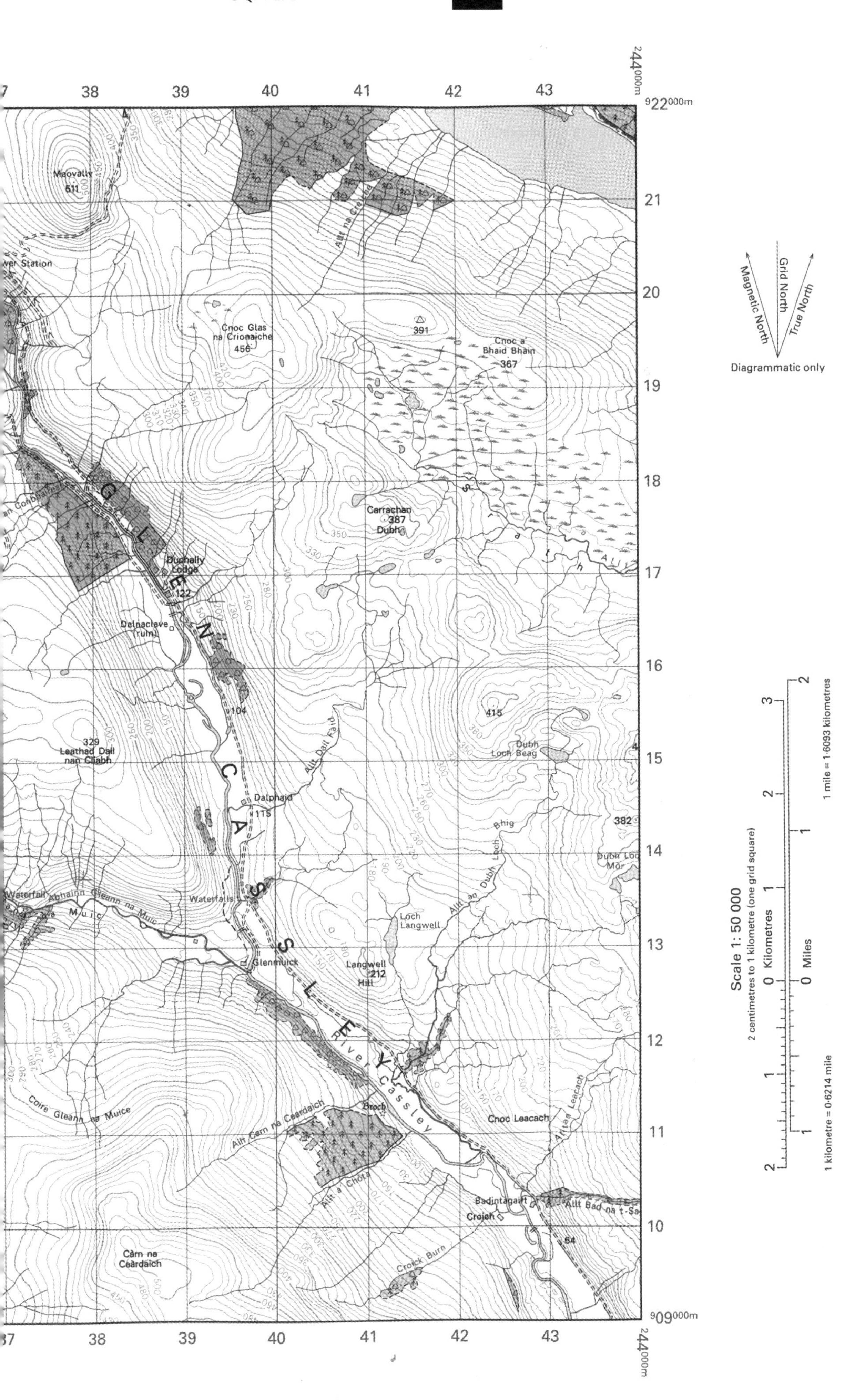

Ordnance Survey, OS, the OS Symbol and Landranger are registered trademarks of Ordnance Survey, the national mapping agency of Great Britain. Reproduction in whole or in part by any means is prohibited without the prior written permission of Ordnance Survey. **For educational use only.**

MARKS

SECTION 1 — PHYSICAL ENVIRONMENTS — 20 marks

Attempt EITHER Question 1 or Question 2
AND Questions 3 and 4

Question 1 — Glaciated Uplands

(a) Study the Ordnance Survey Map Extract (No 2072/15 & 16).

Match these glaciated features with the correct grid references

Features: **U-shaped valley; corrie; arête**

Choose from grid references: 354135, 309201, 323143, 326168. 3

(b) **Explain** the formation of a corrie.

You may use a diagram(s) in your answer. 4

Total marks 7

NOW ANSWER QUESTIONS 3 AND 4

DO NOT ANSWER THIS QUESTION IF YOU HAVE ALREADY ANSWERED QUESTION 1

Question 2 — River and Valleys

(a) Study the Ordnance Survey Map Extract (No 2072/15 & 16).

Match these river features with the correct grid references

Features: **ox-bow lake; meander; V-shaped valley**

Choose from grid references: 389151, 297207, 427099, 423107. 3

(b) **Explain** the formation of a waterfall.

You may use a diagram(s) in your answer. 4

Total marks 7

NOW ANSWER QUESTIONS 3 AND 4

MARKS

Question 3

Diagram Q3 — Land Uses

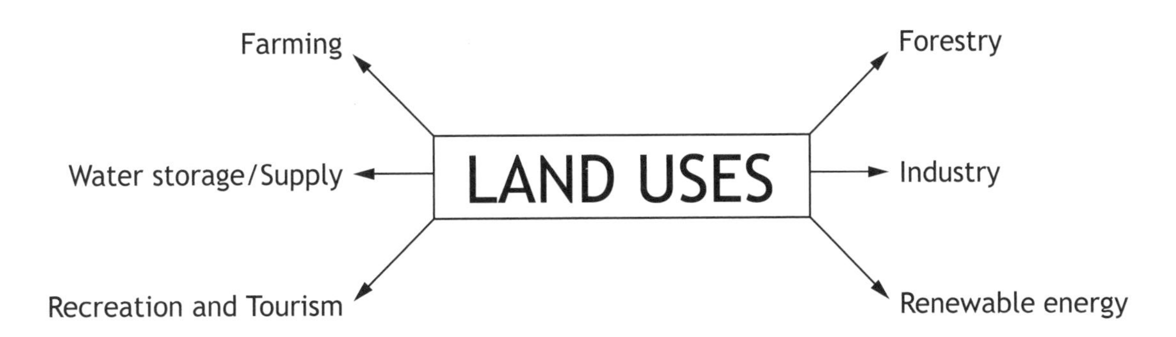

Look at Diagram Q3 and the whole of the OS Map extract.

For **one** of the land uses shown give the advantages of this area for your chosen land use. You must use map evidence. **5**

[Turn over

MARKS

Question 4

Diagram Q4A — Air Masses affecting the British Isles.

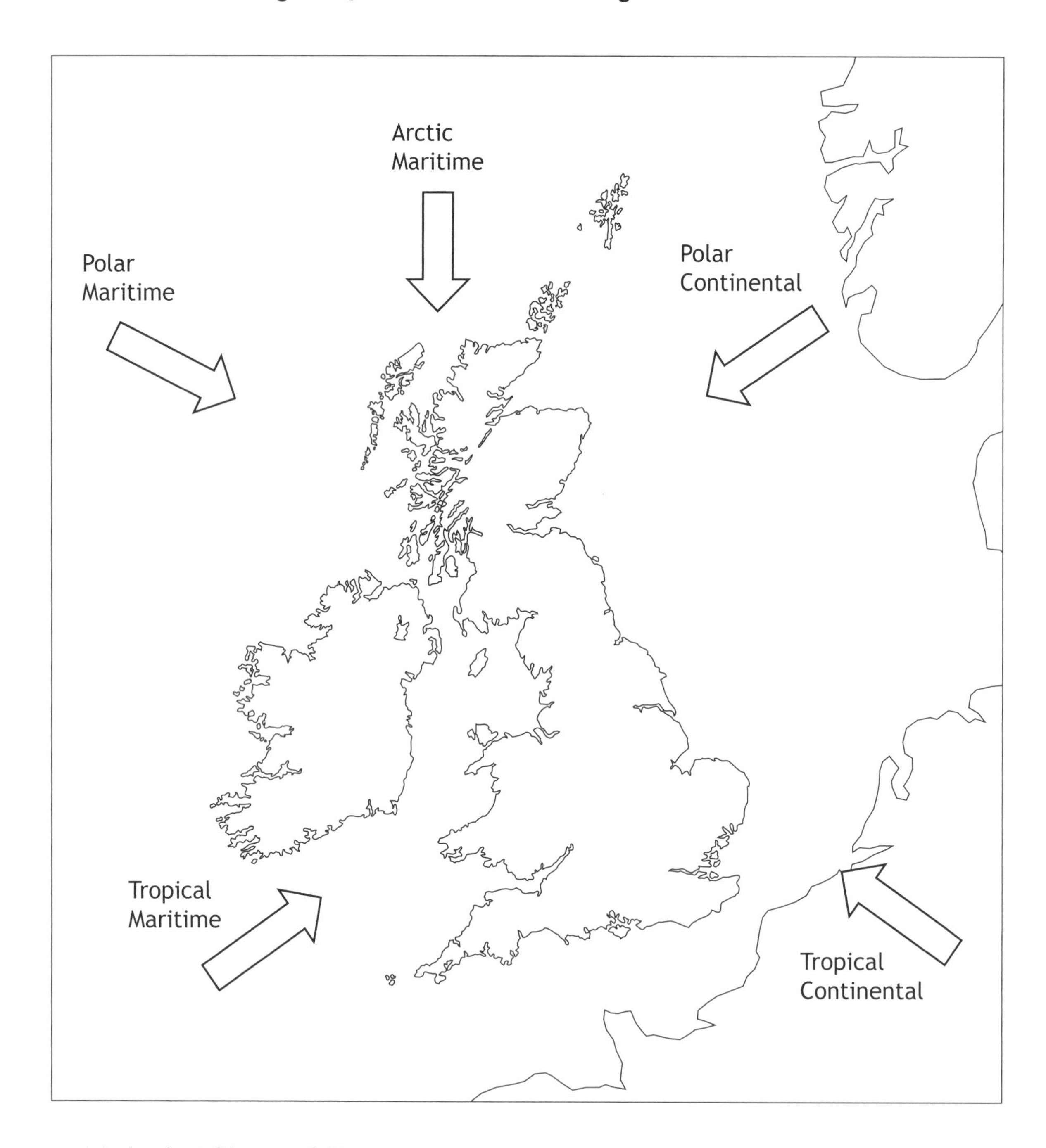

(a) Look at Diagram Q4A.

Describe how a prolonged spell with a **tropical continental** air mass in **summer** would affect the people of the British Isles. 3

MARKS

Diagram Q4B — Synoptic Chart for March 2012

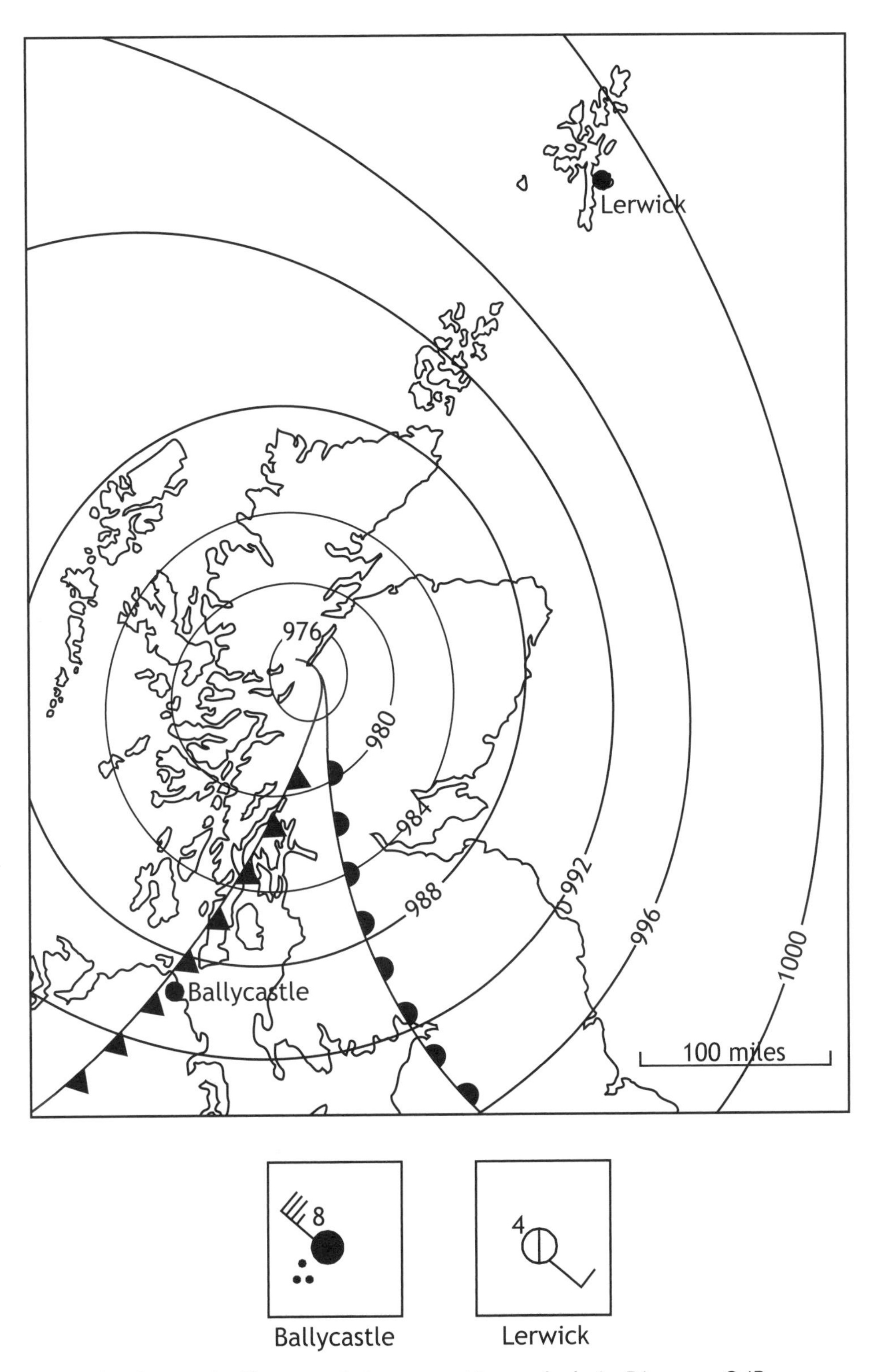

(b) Study the Synoptic Chart and the synoptic symbols in Diagram Q4B.

Give reasons for the **differences** in the weather conditions between Ballycastle and Lerwick. **5**

Total marks 8

MARKS

SECTION 2 — HUMAN ENVIRONMENTS — 20 marks

Attempt Questions 5, 6, 7 and 8

Question 5

Diagram Q5 — Changes in Glasgow's CBD

Expensive Designer Stores	Old Warehouses Converted Into Flats	Indoor Shopping Malls

Look at Diagram Q5.

For Glasgow, or any other developed world city you have studied, **explain** the main changes which have taken place in the CBD over recent years. **5**

MARKS

Question 6

Diagram Q6 — Age Groups in China 1982 to 2050 (Projected)

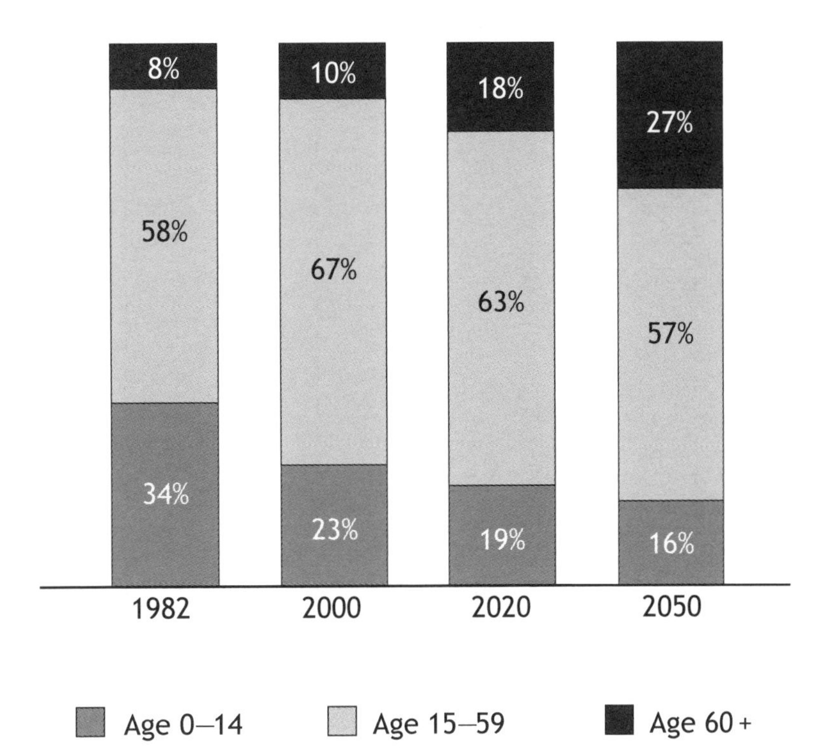

Study Diagram Q6.

(a) **Describe**, **in detail**, the changes in China's age groups between 1982 and 2050 (projected). 4

(b) For China, or other countries you have studied, **describe** methods which have been used to reduce population growth. 4

Total marks 8

[Turn over

MARKS

Question 7

Diagram Q7 — Population Trends in Mumbai and Glasgow

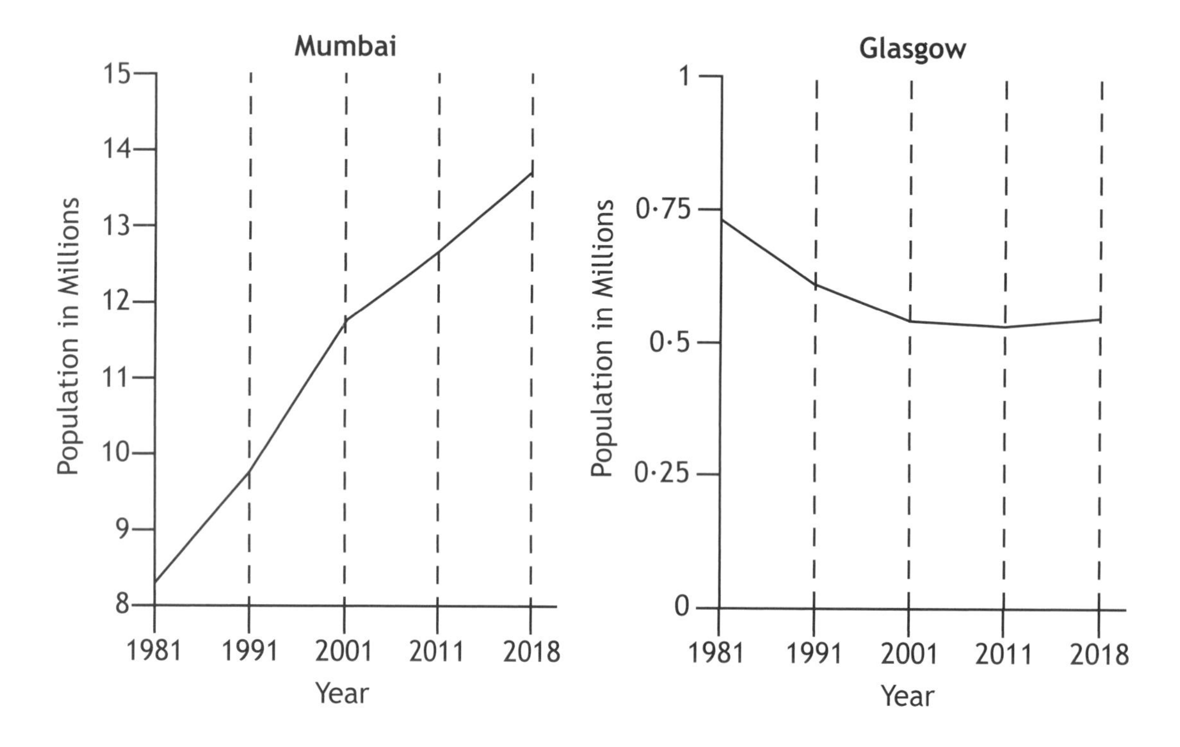

Study Diagram Q7.

Describe, **in detail**, differences in population trends between Mumbai and Glasgow. 3

MARKS

Question 8

Diagram Q8 — Developments in Farming

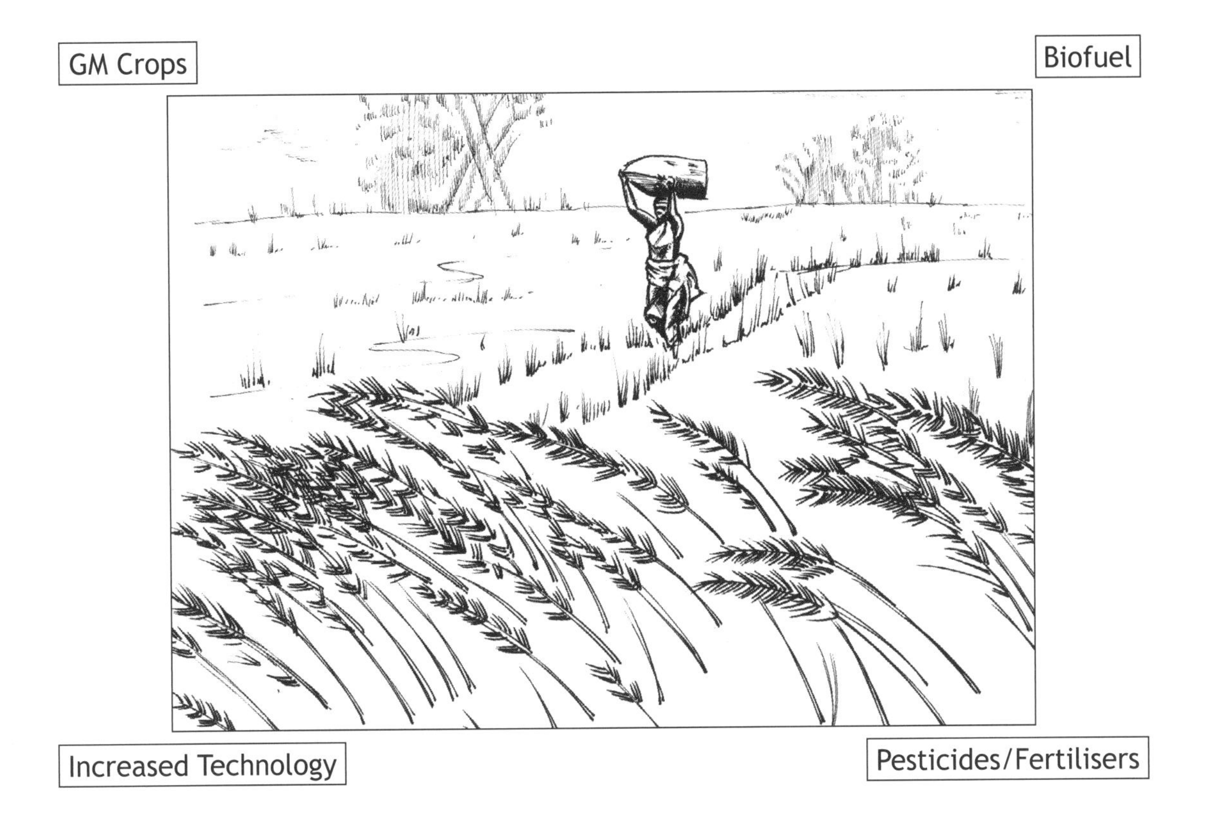

Look at Diagram Q8.

Explain how recent developments in agriculture in developing countries are helping farmers. 4

[Turn over

SECTION 3 — GLOBAL ISSUES — 20 marks

Attempt any TWO questions

MARKS

Question 9 — Climate Change

Diagram Q9A — Areas at greater Risk from Climate Change

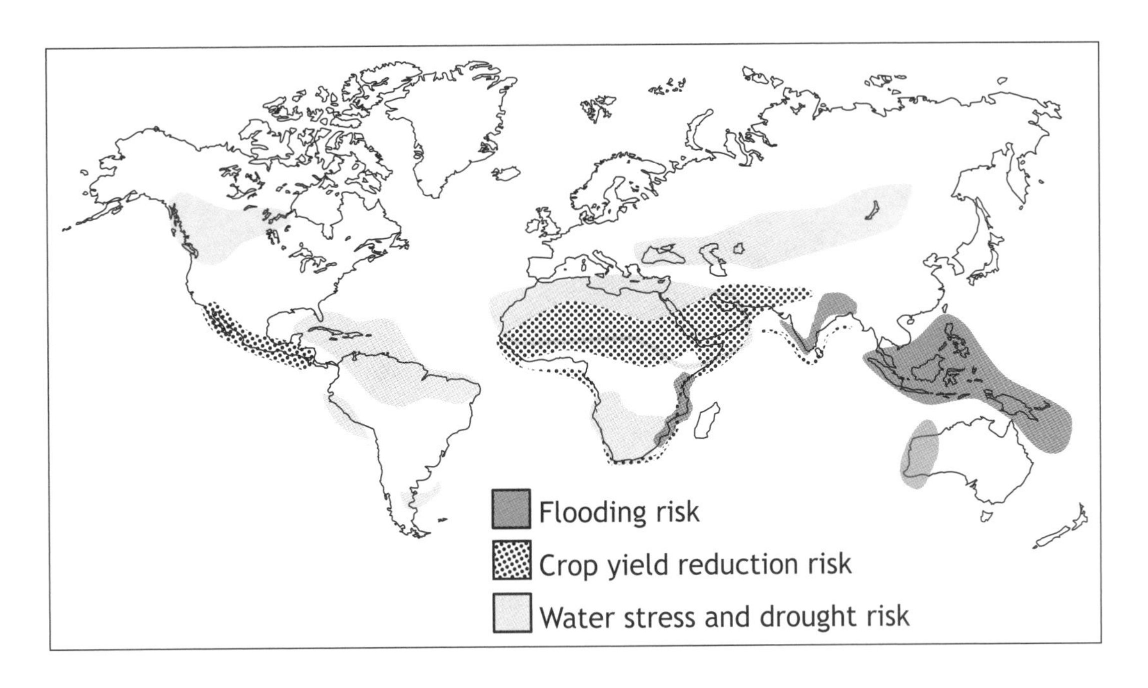

(a) Study Diagram Q9A.

Describe, **in detail**, the locations which are at greater risk from climate change. 4

Diagram Q9B — Evidence of Climate Change

(b) Look at Diagram Q9B.

Explain, **in detail**, the physical **and** human causes of global climate change. 6

Total marks 10

[Turn over

MARKS

QUESTION 10 — Impact of Human Activity on the Natural Environment

Diagram Q10A — Barrow Alaska

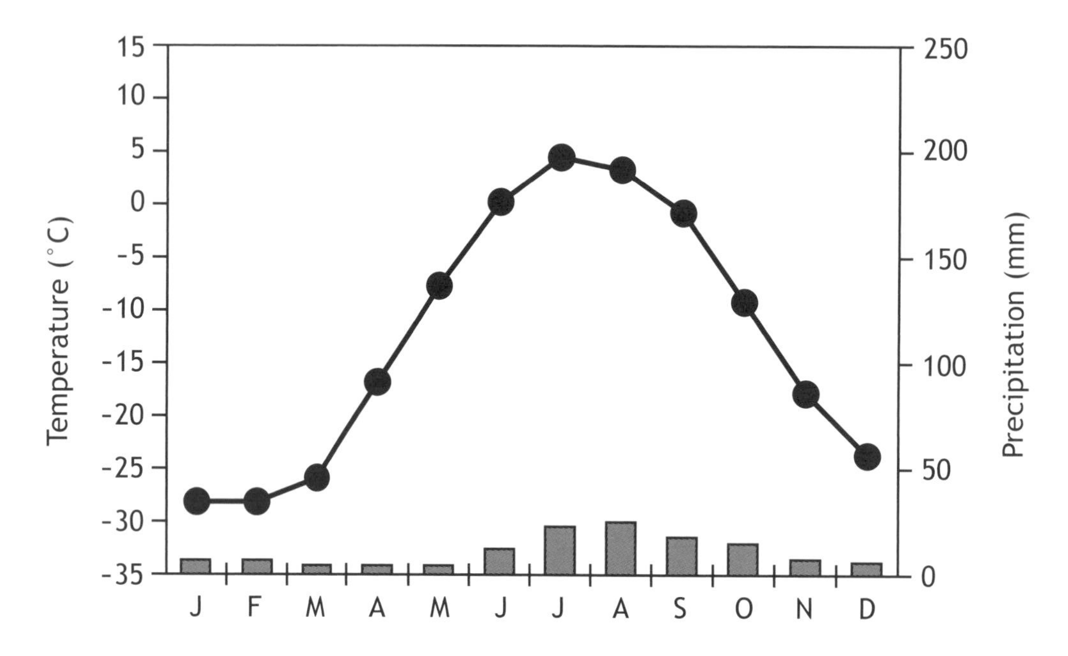

Diagram Q10B — Location of Tundra Climatic Area

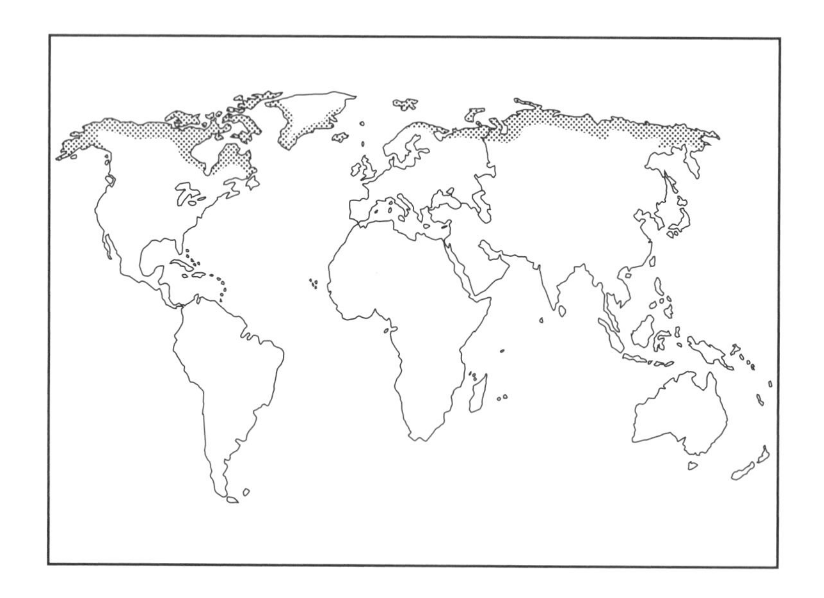

(a) Study Diagrams Q10A and Q10B.

Describe, **in detail**, the main features of the Tundra climate. 4

MARKS

Diagram Q10C — Impact of Human Activities in the Tundra

(b) Look at Diagram Q10C.

Describe the advantages **and** disadvantages brought about by human activities in the Tundra.

You may refer to an area you have studied in your answer. **6**

Total marks 10

[Turn over

MARKS

Question 11 — Environmental Hazards

Diagram Q11A — Distribution of Volcanoes

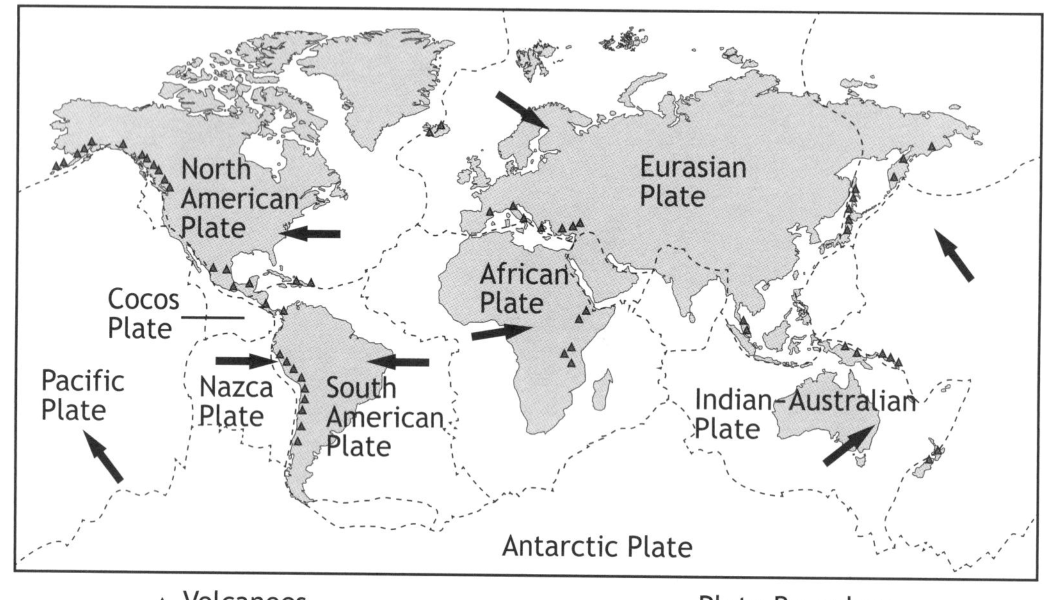

(a) Study Diagram Q11A.

Describe, in detail, the distribution of the world's volcanoes. 4

MARKS

Diagram Q11B — Japanese Earthquake, 2011

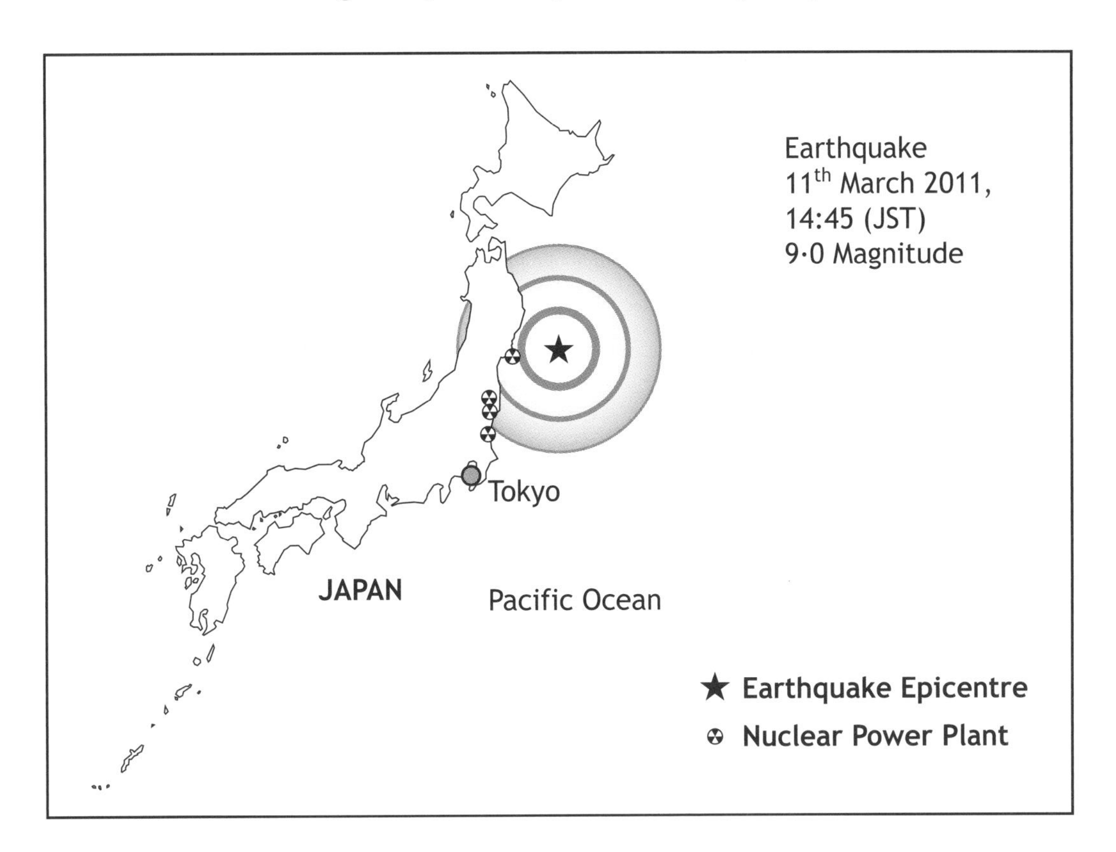

(b) Look at Diagram Q11B.

For the Japanese earthquake, or any other named earthquake you have studied, **describe**, **in detail**, the impact on people **and** the landscape. 6

Total marks 10

[Turn over

MARKS

Question 12 — Trade and Globalisation

Diagram Q12A — World Exports by Region

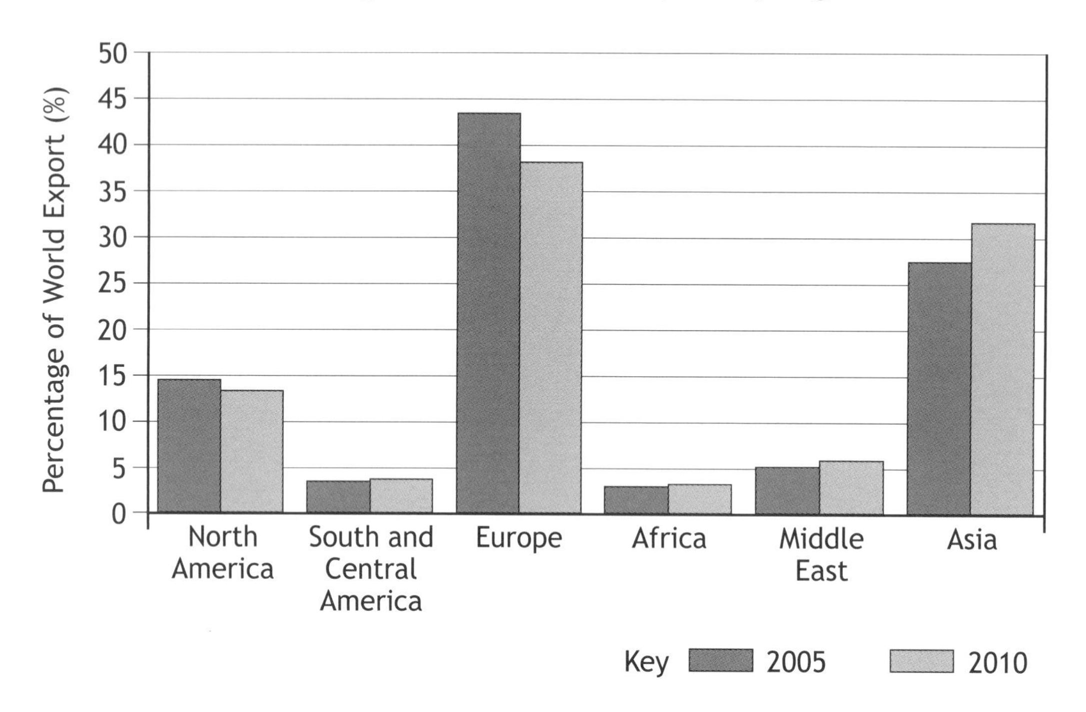

(a) Study Diagram Q12A.

Describe, **in detail**, the change in World Exports from 2005 to 2010. 4

MARKS

Diagram Q12B — Collecting Fairtrade Coffee Beans

(b) Look at Diagram Q12B.

Explain how buying Fairtrade products helps people in the developing world. 6

Total marks 10

[Turn over

MARKS

Question 13 — Tourism

Diagram Q13A — Top Ten World Tourist Destinations (millions of visitors per year)

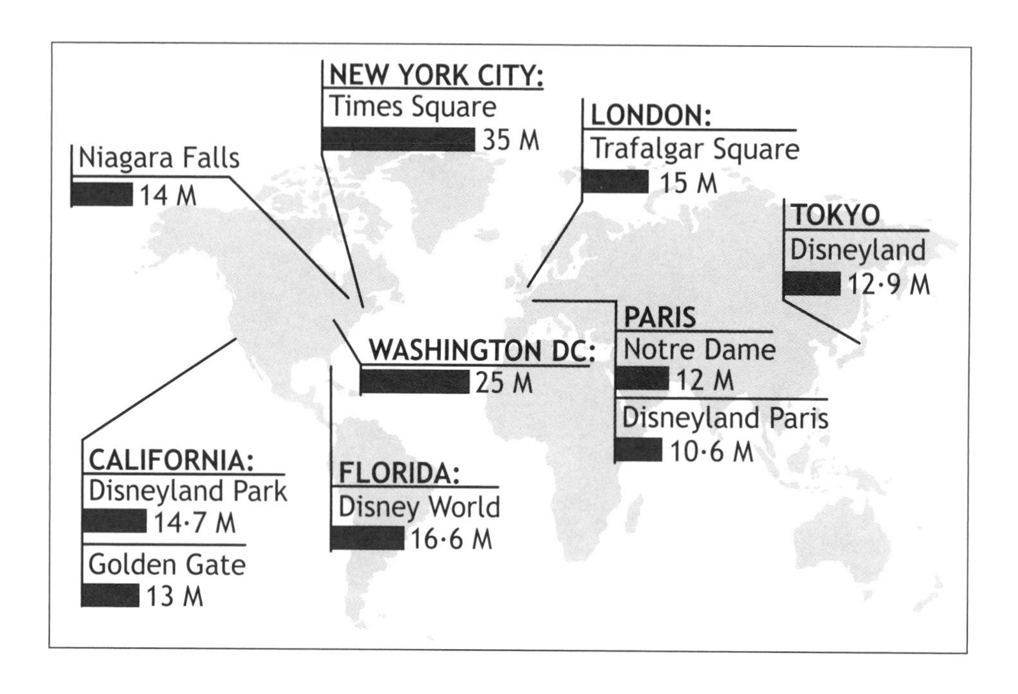

(a) Study Diagram Q13A.

Describe, **in detail**, the distribution of the top ten world tourist destinations. 4

MARKS

Diagram Q13B — Mass Tourism on an Italian Beach

(b) Look at Diagram Q13B.

Describe the effects of mass tourism on people **and** the environment. 6

Total marks 10

[Turn over

MARKS

Question 14 — Health

Diagram Q14A — Worldwide Child Deaths under 5 Years (per 1,000 live births)

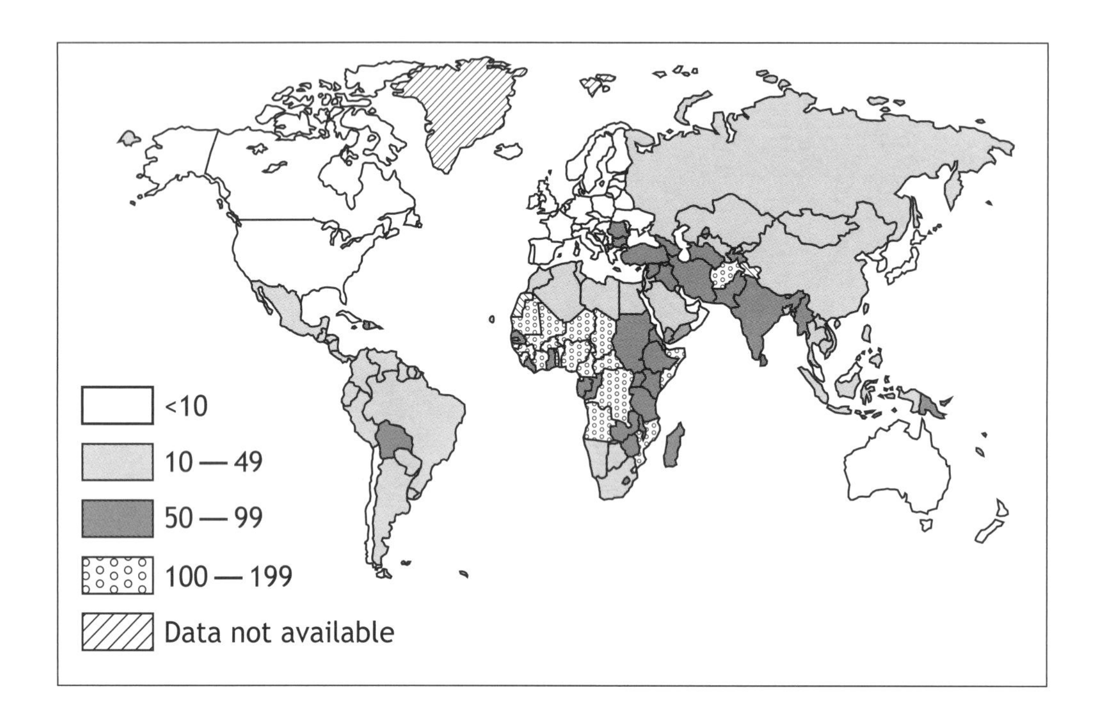

(a) Study Diagram Q14A.

Describe, **in detail**, the distribution of worldwide child deaths under the age of 5. 4

MARKS

Diagram Q14B — World AIDS Day Report 2012

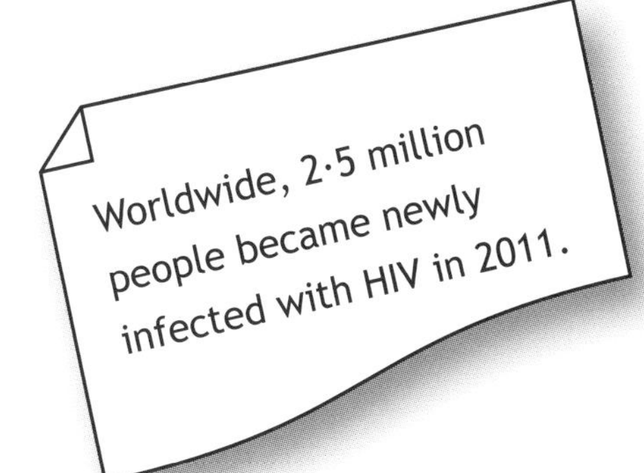

The number of people accessing HIV treatment increased by 63% from 2009 to 2011.

(b) Look at Diagram Q14B.

Explain methods used to limit the spread of AIDS in developed **and** developing countries. **6**

Total marks 10

[END OF QUESTION PAPER]

[BLANK PAGE]

DO NOT WRITE ON THIS PAGE

NATIONAL 5

2015

X733/75/11 **Geography**

THURSDAY, 21 MAY

9:00 AM – 10:45 AM

Total marks — 60

SECTION 1 — PHYSICAL ENVIRONMENTS — 20 marks

Attempt EITHER question 1 **OR** question 2. **ALSO** attempt questions 3, 4 and 5.

SECTION 2 — HUMAN ENVIRONMENTS — 20 marks

Attempt questions 6, 7 and 8

SECTION 3 — GLOBAL ISSUES — 20 marks

Attempt any TWO of the following

Question 9 — Climate Change
Question 10 — Impact of Human Activity on the Natural Environment
Question 11 — Environmental Hazards
Question 12 — Trade and Globalisation
Question 13 — Tourism
Question 14 — Health

Credit will always be given for appropriately labelled sketch maps and diagrams.

Write your answers clearly in the answer booklet provided. In the answer booklet you must clearly identify the question number you are attempting.

Use **blue** or **black** ink.

Before leaving the examination room you must give your answer booklet to the Invigilator; if you do not, you may lose all the marks for this paper.

MARKS

SECTION 1 — PHYSICAL ENVIRONMENTS — 20 marks

Attempt EITHER Question 1 or Question 2
AND Questions 3, 4 and 5

Question 1 — Coastal Landscapes

Study the Ordnance Survey Map Extract (Item A) of the Salcombe area.

(a) Match these grid references with the correct coastal features

Grid references: **766356**, **674398**, **690382**

Choose from features: cliff; headland; bay; stack. 3

(b) **Explain** the formation of **one** of the coastal features listed in part (a).

You may use a diagram(s) in your answer. 4

NOW ANSWER QUESTIONS 3, 4 AND 5

DO NOT ANSWER THIS QUESTION IF YOU HAVE ALREADY ANSWERED QUESTION 1

Question 2 — Rivers and Valleys

Study the Ordnance Survey Map Extract (Item A) of the Salcombe area.

(a) Match these grid references with the correct river features

Grid references: **708473**, **713410**, **684466**

Choose from features: levée; meander; v-shaped valley; waterfall. 3

(b) **Explain** the formation of **one** of the river features listed in part (a).

You may use a diagram(s) in your answer. 4

NOW ANSWER QUESTIONS 3, 4 AND 5

MARKS

Question 3

Diagram Q3: Quote from a Local Landowner

"This area has the potential for a variety of different land uses, including farming, forestry, recreation/tourism, water storage/supply, industry and renewable energy."

Study Diagram Q3 and the Ordnance Survey Map Extract (Item A) of the Salcombe area.

Choose **two** different land uses shown in Diagram Q3.

Using map evidence, **explain** how the area shown on the map extract is suitable for your chosen land uses. 5

Question 4

Diagram Q4: Selected Land Uses

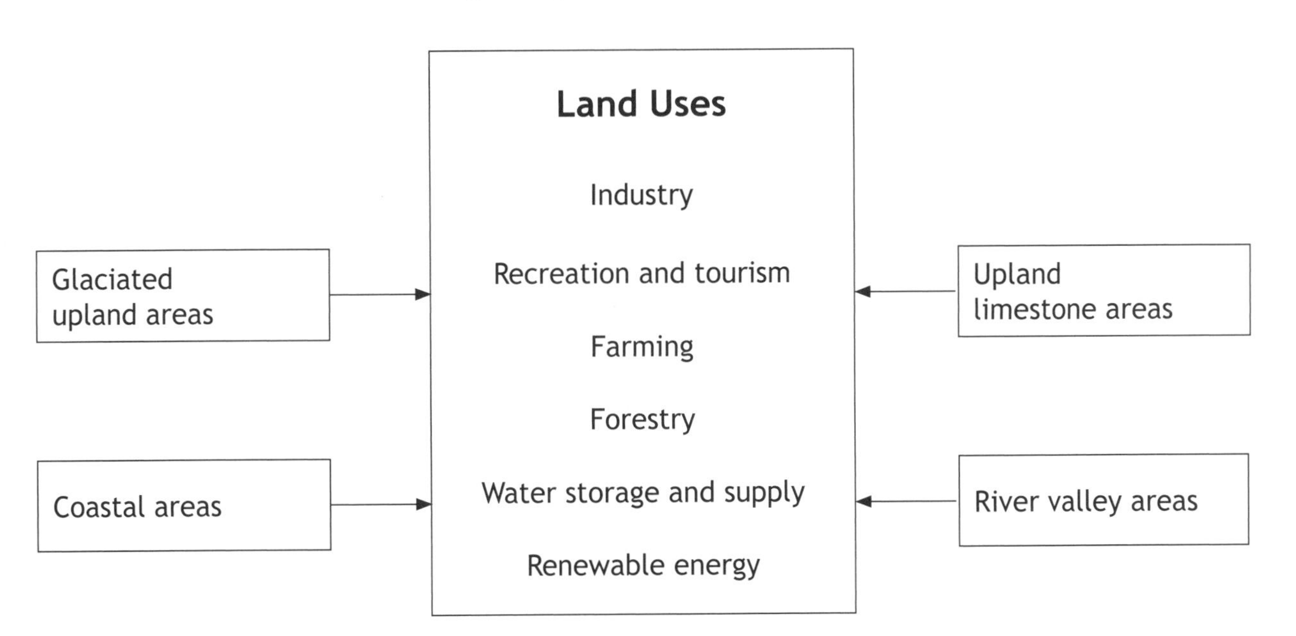

Look at Diagram Q4 above.

For a named area you have studied, **explain**, **in detail**, ways in which **two** different land uses may be in conflict with each other. 4

[Turn over

MARKS

Question 5

Diagram Q5: Average UK Temperatures in July

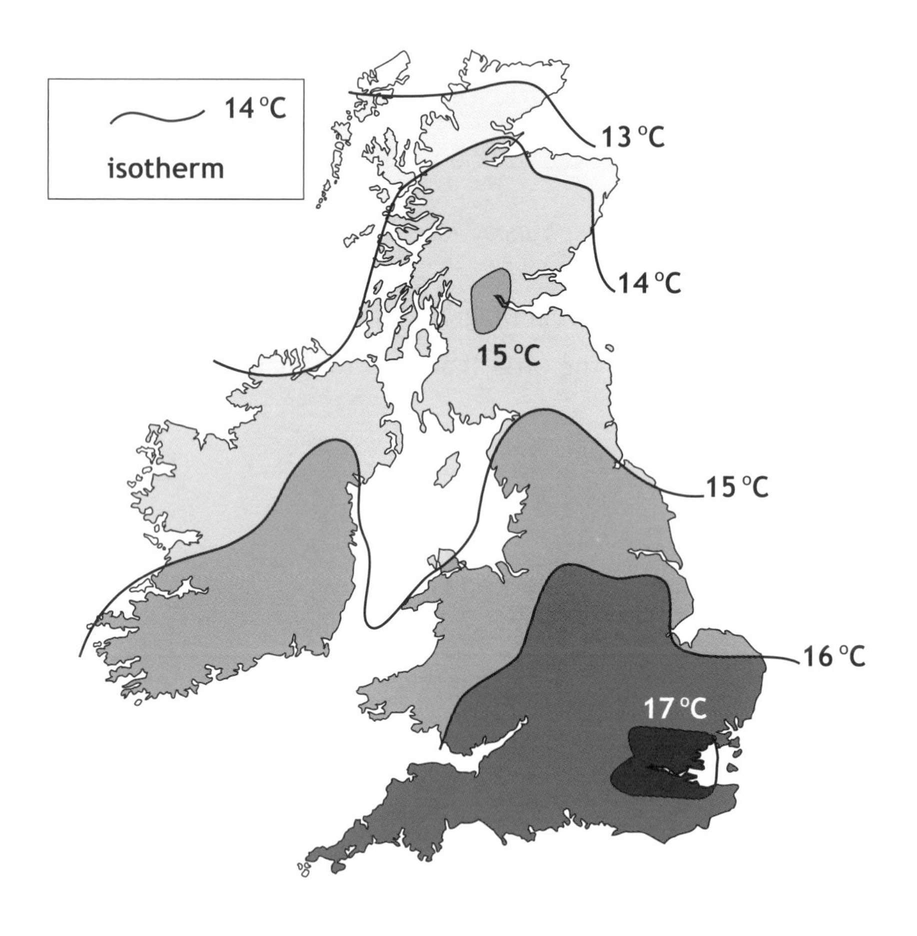

Look at Diagram Q5.

Explain the factors which cause differences in average UK temperatures. 4

MARKS

SECTION 2 — HUMAN ENVIRONMENTS — 20 marks

Attempt Questions 6, 7 and 8

Question 6

Diagram Q6

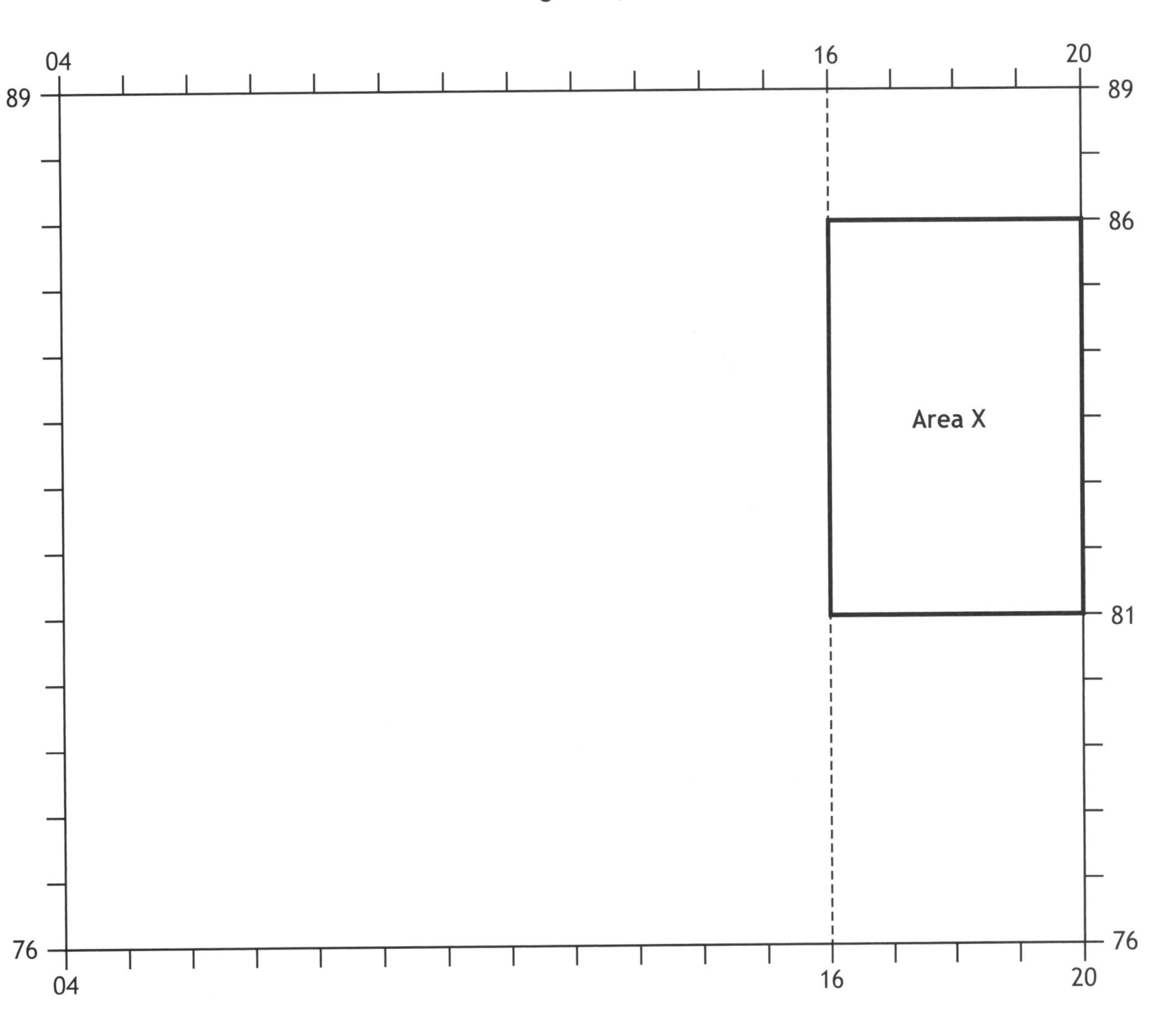

Study the Ordnance Survey Map Extract (Item B) of the Birmingham area and Diagram Q6 above.

(a) Give map evidence to show that part of the Central Business District (CBD) of Birmingham is found in grid square 0786. **3**

(b) Find Area X on Diagram Q6 and the map extract (Item B).

Birmingham Airport, a golf course, a business park and a housing area are found in Area X on the rural/urban fringe of Birmingham. Using map evidence **explain** why such developments are found there. **5**

MARKS

Question 7

Diagram Q7: Births in Scotland 1901–2011

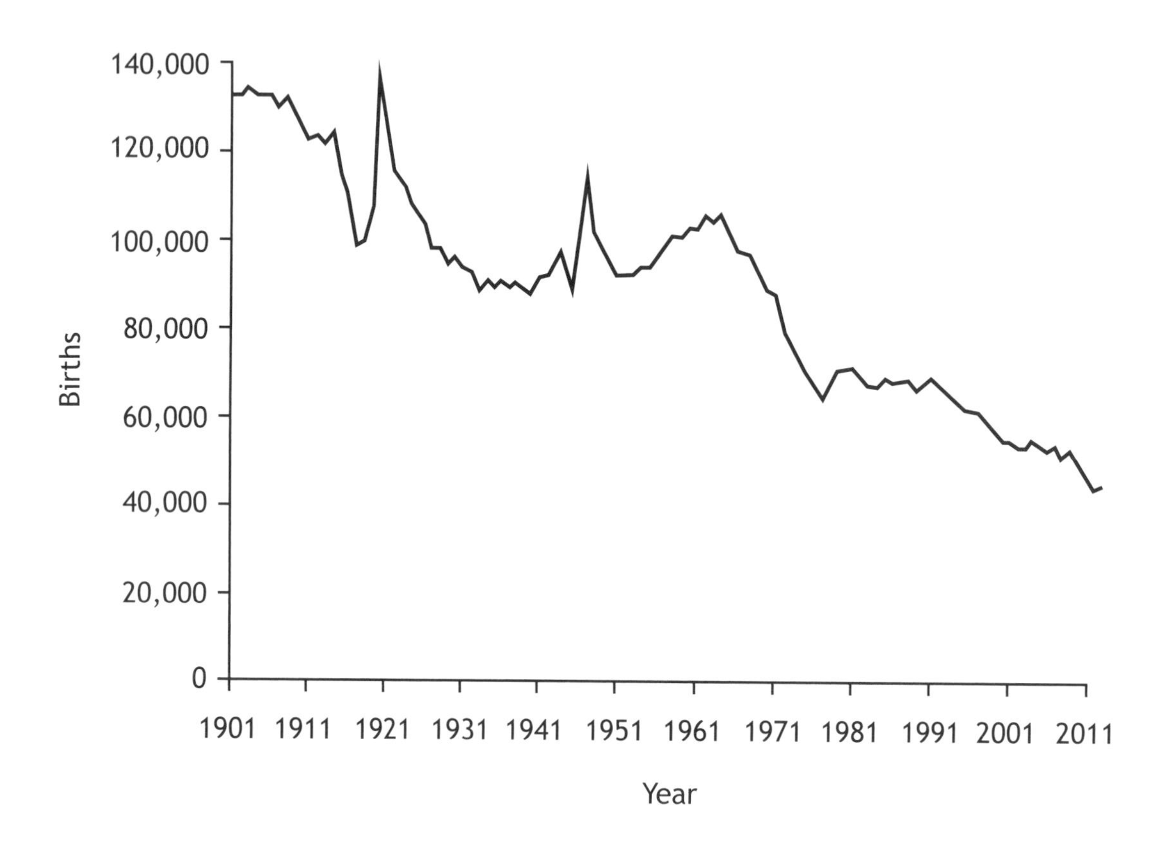

Look at Diagram Q7

Give reasons why the birth rate has decreased in developed countries such as Scotland. **6**

MARKS

Question 8

Diagram Q8: Shanty Town Improvements in Brazil

Look at Diagram Q8

For a named city in the developing world **describe**, **in detail**, measures taken to improve conditions in shanty towns. 6

[Turn over

SECTION 3 — GLOBAL ISSUES — 20 marks

Attempt any TWO questions

MARKS

Question 9 — Climate Change

Diagram Q9: Area of Arctic Sea Ice (1979–2013)

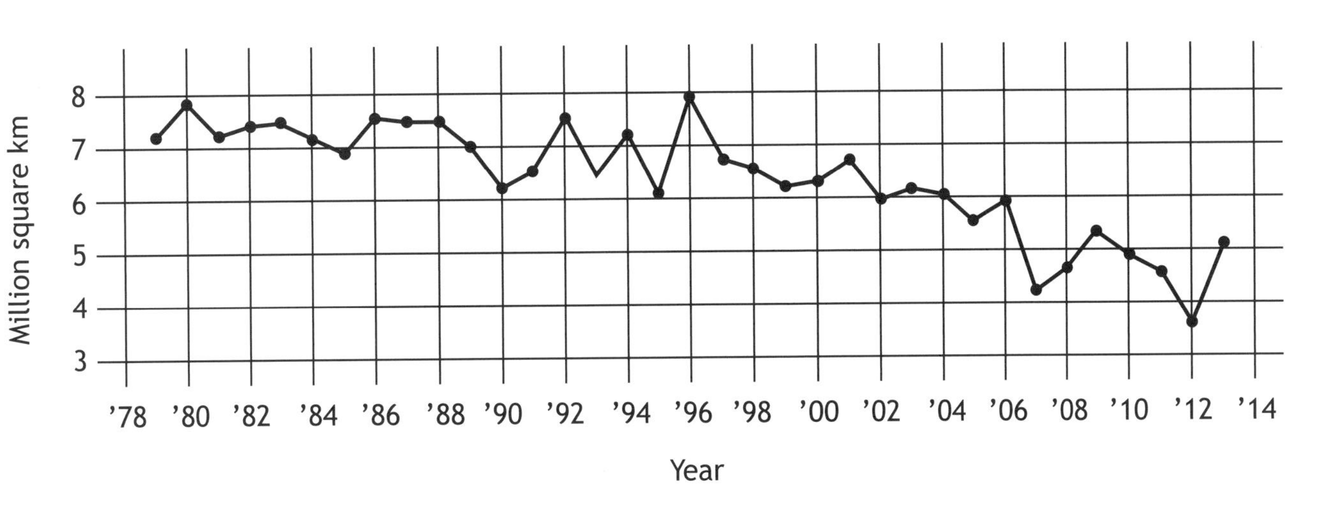

Study Diagram Q9

(a) **Describe**, **in detail**, the changes in the area of Arctic Sea ice. **4**

(b) Melting sea ice is one effect of climate change.

Explain some other effects of climate change. **6**

[Turn over

MARKS

Question 10 — Impact of Human Activity on the Natural Environment

Diagram Q10A: Deforestation in Peru 2004–2012

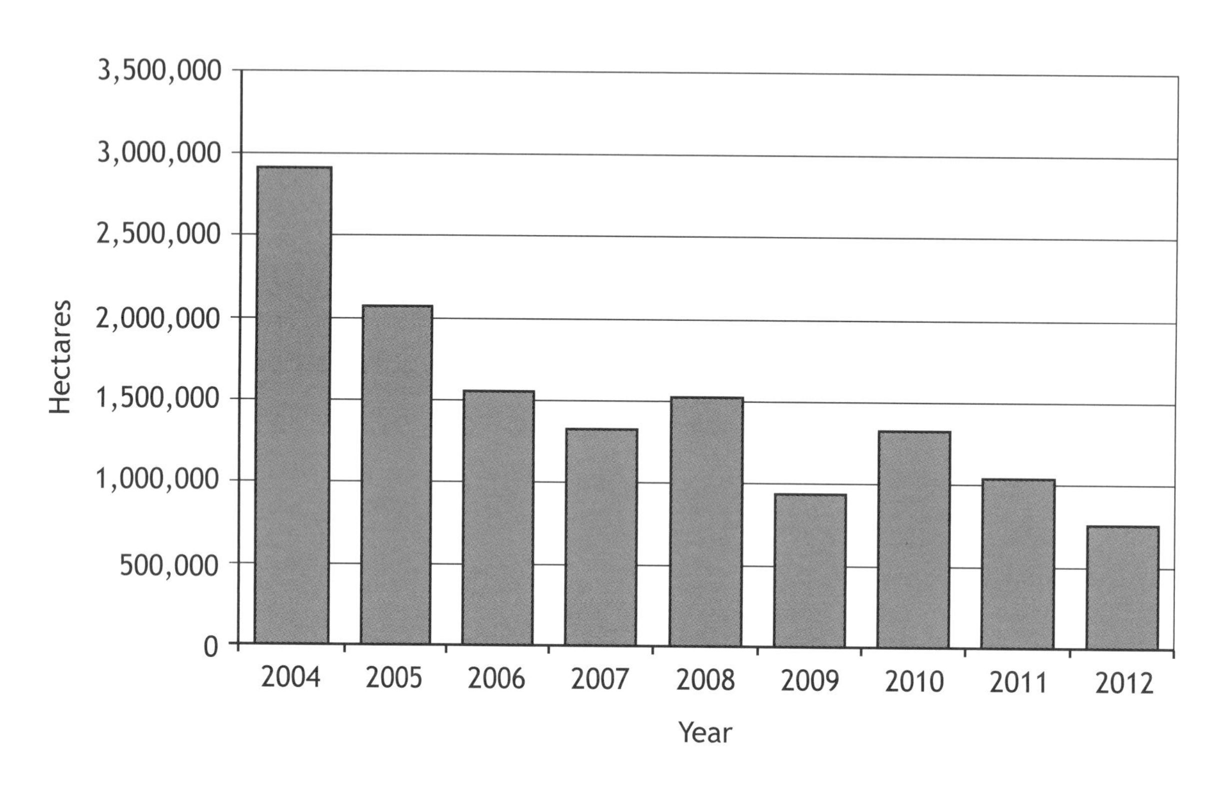

(a) Study Diagram Q10A.

Describe, **in detail**, the changes in deforestation in Peru from 2004 to 2012. 4

Diagram Q10B: Human Activity in the Tundra and Equatorial regions.

Oil pipeline in the Tundra

Cattle ranching in the Rainforest

(b) Look at Diagram Q10B.

For a named area you have studied, **explain** the impact of recent human activity on people and the environment. 6

MARKS

Question 11 — Environmental Hazards

Diagram Q11A: Earthquake Threatened Cities

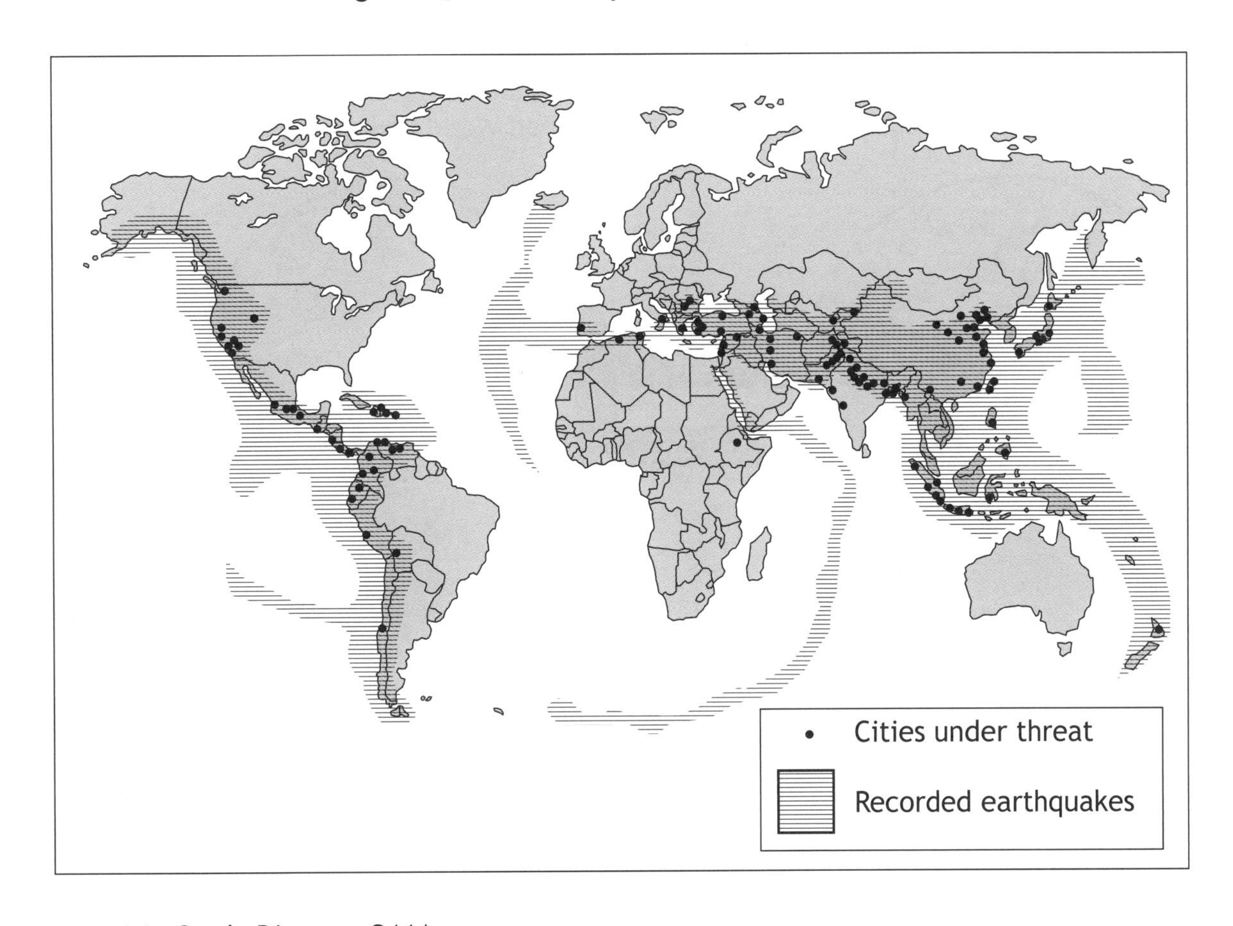

(a) Study Diagram Q11A.

Describe, **in detail**, the distribution of cities most threatened by earthquakes. 4

Diagram Q11B

(b) Look at Diagram Q11B.

Explain, **in detail**, the strategies used to reduce the impact of an earthquake.

You must refer to named examples you have studied in your answer. 6

Question 12 — Trade and Globalisation **MARKS**

Diagram Q12A: Pattern of World Trade

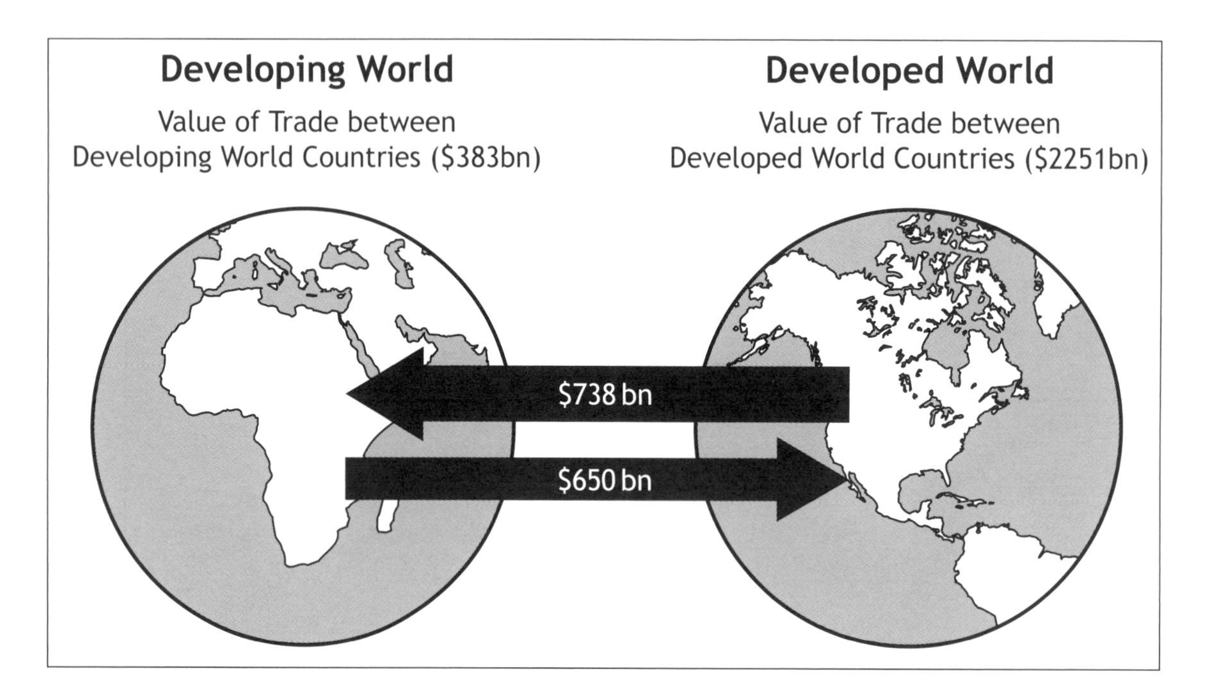

(a) Study Diagram Q12A.

Describe, **in detail**, the pattern of world trade. 4

Diagram Q12B: Trade Between Africa and the European Union (EU)

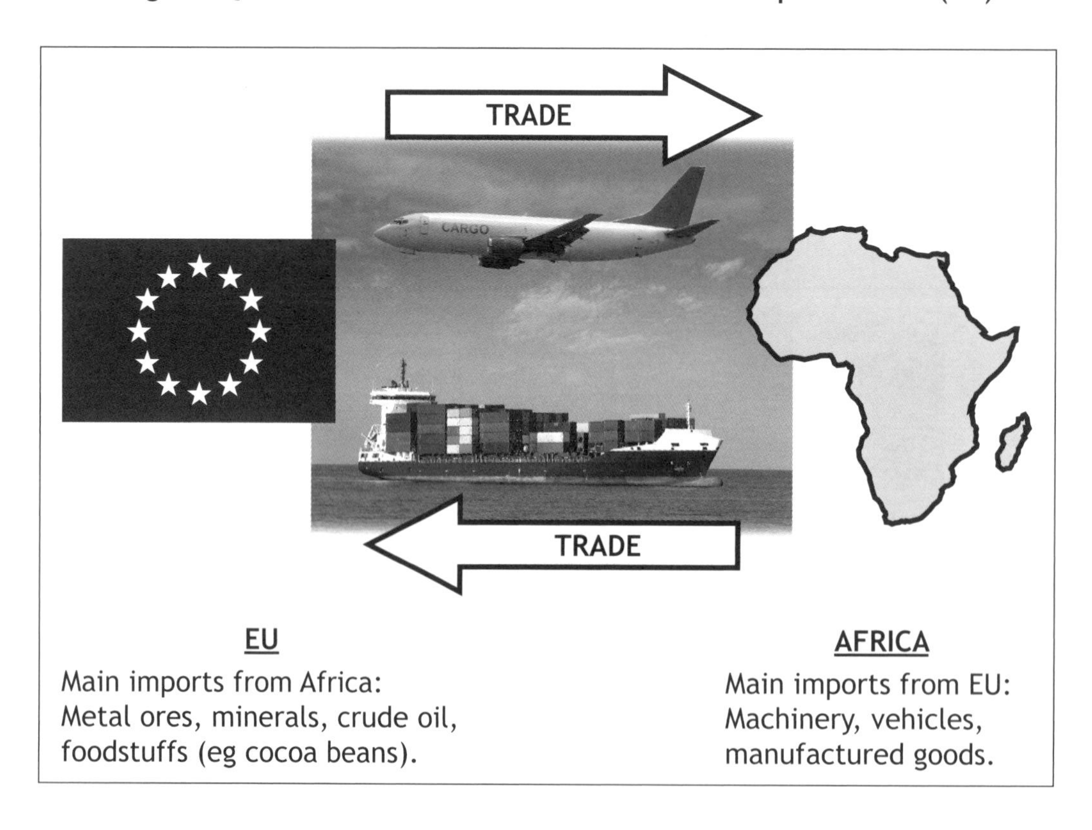

(b) Look at Diagram Q12B.

Referring to example(s) you have studied, **describe** the impact of world trade on people and the environment. 6

MARKS

Question 13 — Tourism

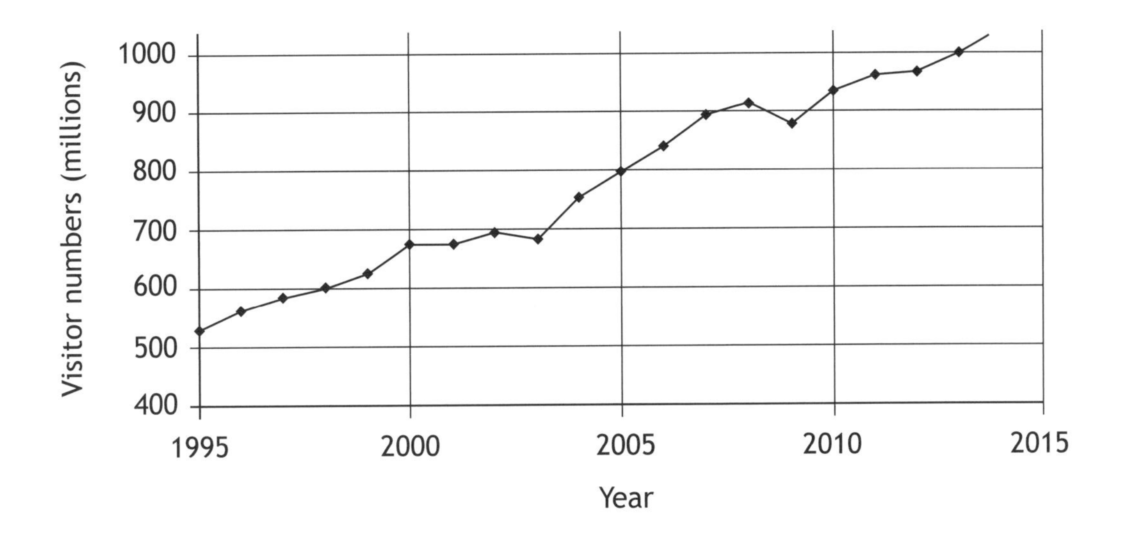

(a) Study Diagram Q13A.

Describe, **in detail**, the changes in global visitor numbers since 1995. **4**

Diagram Q13B: Quote from Rainforest Community Leader

> "ECO-TOURISM has helped us to support environmental protection and improve the well-being of our people all year round."

(b) Look at Diagram Q13B.

For a named tourist area you have studied, **describe**, **in detail** the impact of eco-tourism on people **and** the environment. **6**

[Turn over for Question 14 on *Page fourteen*

MARKS

Question 14 — Health

Diagram Q14A: Worldwide Male Deaths from Heart Disease in 2011 (per 100,000 Males)

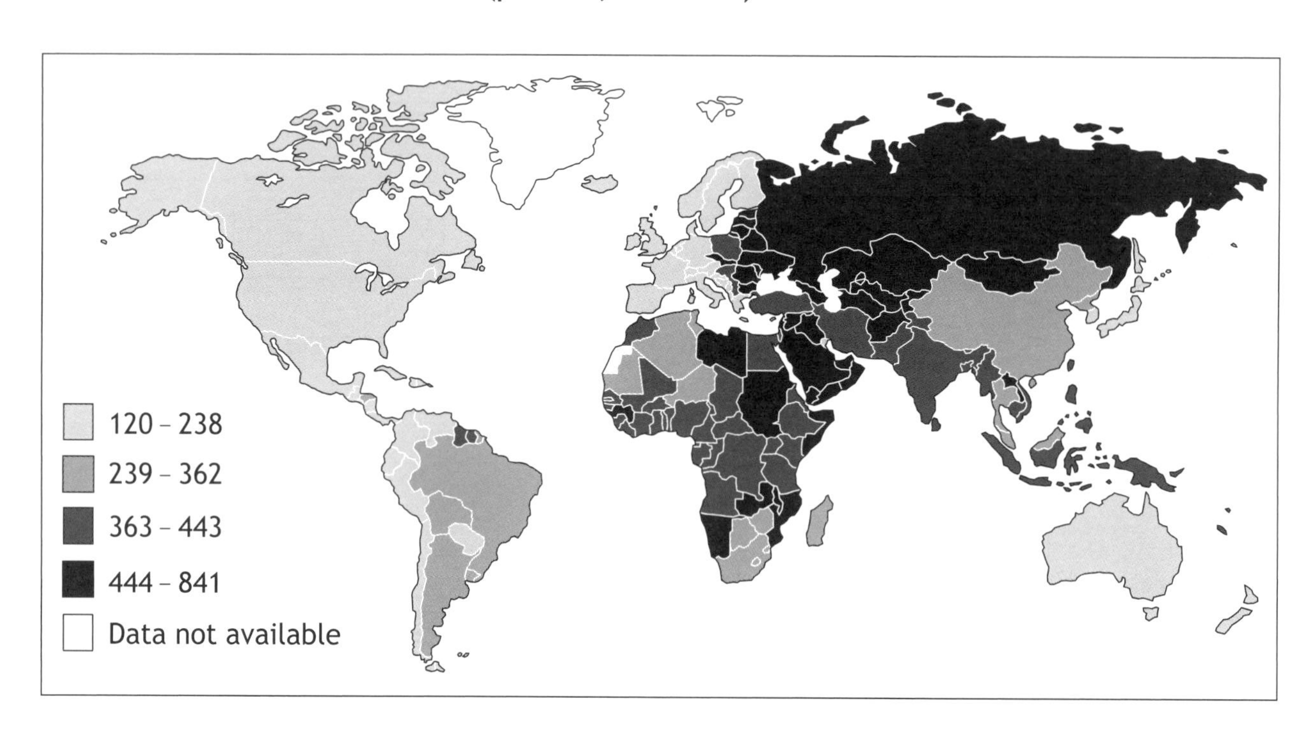

(a) Study Diagram Q14A.

Describe, **in detail**, the distribution of male deaths from heart disease. 4

Diagram Q14B: Selected Developing World Diseases

- **Malaria**
- **Cholera**
- **Kwashiorkor**
- **Pneumonia**

(b) Choose **one** disease from Diagram Q14B above.

For the disease you have chosen, **explain** the methods used to control it. 6

[END OF QUESTION PAPER]

[BLANK PAGE]

DO NOT WRITE ON THIS PAGE

[BLANK PAGE]

DO NOT WRITE ON THIS PAGE

X733/75/21

Geography
Ordnance Survey Map
Item A

THURSDAY, 21 MAY

9:00 AM – 10:45 AM

The colours used in the printing of these map extracts are indicated in the four little boxes at the top of the map extract. Each box should contain a colour; if any does not, the map is incomplete and should be returned to the Invigilator.

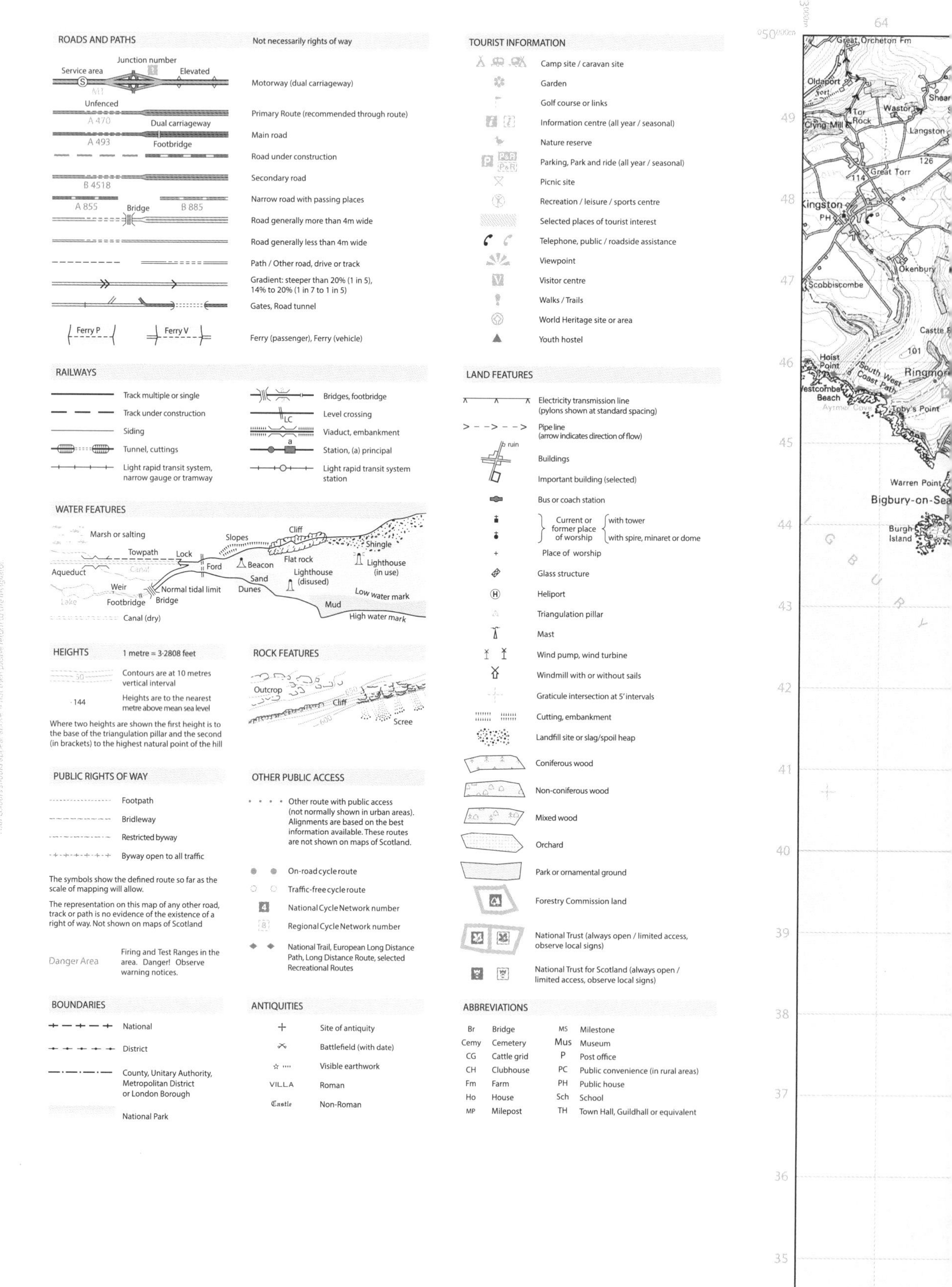
ROADS AND PATHS
Not necessarily rights of way
Motorway (dual carriageway)
Primary Route (recommended through route)
Main road
Road under construction
Secondary road
Narrow road with passing places
Road generally more than 4m wide
Road generally less than 4m wide
Path / Other road, drive or track
Gradient: steeper than 20% (1 in 5), 14% to 20% (1 in 7 to 1 in 5)
Gates, Road tunnel
Ferry (passenger), Ferry (vehicle)
RAILWAYS
Track multiple or single
Track under construction
Siding
Tunnel, cuttings
Light rapid transit system, narrow gauge or tramway
Bridges, footbridge
Level crossing
Viaduct, embankment
Station, (a) principal
Light rapid transit system station
WATER FEATURES
HEIGHTS
1 metre = 3·2808 feet
Contours are at 10 metres vertical interval
Heights are to the nearest metre above mean sea level
Where two heights are shown the first height is to the base of the triangulation pillar and the second (in brackets) to the highest natural point of the hill
ROCK FEATURES
PUBLIC RIGHTS OF WAY
Footpath
Bridleway
Restricted byway
Byway open to all traffic
The symbols show the defined route so far as the scale of mapping will allow.
The representation on this map of any other road, track or path is no evidence of the existence of a right of way. Not shown on maps of Scotland
Firing and Test Ranges in the area. Danger! Observe warning notices.
OTHER PUBLIC ACCESS
Other route with public access (not normally shown in urban areas). Alignments are based on the best information available. These routes are not shown on maps of Scotland.
On-road cycle route
Traffic-free cycle route
National Cycle Network number
Regional Cycle Network number
National Trail, European Long Distance Path, Long Distance Route, selected Recreational Routes
BOUNDARIES
National
District
County, Unitary Authority, Metropolitan District or London Borough
National Park
ANTIQUITIES
Site of antiquity
Battlefield (with date)
Visible earthwork
Roman
Non-Roman
TOURIST INFORMATION
Camp site / caravan site
Garden
Golf course or links
Information centre (all year / seasonal)
Nature reserve
Parking, Park and ride (all year / seasonal)
Picnic site
Recreation / leisure / sports centre
Selected places of tourist interest
Telephone, public / roadside assistance
Viewpoint
Visitor centre
Walks / Trails
World Heritage site or area
Youth hostel
LAND FEATURES
Electricity transmission line (pylons shown at standard spacing)
Pipe line (arrow indicates direction of flow)
Buildings
Important building (selected)
Bus or coach station
Place of worship
Glass structure
Heliport
Triangulation pillar
Mast
Wind pump, wind turbine
Windmill with or without sails
Graticule intersection at 5' intervals
Cutting, embankment
Landfill site or slag/spoil heap
Coniferous wood
Non-coniferous wood
Mixed wood
Orchard
Park or ornamental ground
Forestry Commission land
National Trust (always open / limited access, observe local signs)
National Trust for Scotland (always open / limited access, observe local signs)
ABBREVIATIONS
Br Bridge
Cemy Cemetery
CG Cattle grid
CH Clubhouse
Fm Farm
Ho House
MP Milepost
MS Milestone
Mus Museum
P Post office
PC Public convenience (in rural areas)
PH Public house
Sch School
TH Town Hall, Guildhall or equivalent
Extract No 2143/202
1:50 000 Scale
Landranger Series
Four colours should appear above; if not then please return to the invigilator.
Great Orcheton Fm
Oldaport
Clyng Mill
Tor Rock
Langston
Great Torr
Kingston
Scobbiscombe
Okenbury
Hoist Point
South West Coast Path
Westcombe Beach
Ringmore
Toby's Point
Warren Point
Bigbury-on-Sea
Burgh Island

67
68
69
70
71
72
73
74
75
76
Stockadon
Alleron
Fernhill Ho
Reveton
Wood Barton
Morecombe Fm
Yetsonais Fm
Ley
Wakeham
Heath Fm
Chantry
Ham Fm
Grimpstonleigh
Tetwell
Cumery
Yanston Fm
Reads Fm
Woodleigh
The Mounts
Fallapit Ho
Combe Fm
Hingston Borough
Idestone
Loddiswell
Torr Quarry
Challon's Combe
Ashford
Nutcombe Fm
Icy Park
Weeke
Weir
Aveton Gifford
Waterhead
New Br
Knap Mill
Rake
Warcombe
Field Study Centre
Borough
St Ann's Chapel
Bridge
Weir
Knighton
Easton
Bridge End
Venn
Hatch
Leigh Barton
Sorley
Slade
Bigbury Court
Ledstone
Goveton
Bigbury
Nuckwell Fm
Lower Sigdon
Centry
Marrifield
Churchstow
Combe Royal
Coombe Fm
Hotel
Buckland-Tout-Saints
Stadbury Fm
Osborne Newton
Croft
Malston Mill
Slade Fm
Elston
Norton
Hexdown
River Avon
Hospl
North Upton
Worthy
West Redford
Bearscombe
Aunemouth
Buckland Park
Dodbrooke
Bowcombe
Bowden
Whitley
Huxton Cross
Westville
Cockleridge
Clanacombe
KINGSBRIDGE
Duncombe Cross
Heddeswell
High Ho
Coll
Buckland
Upton
West Alvington
Southville
Bantham
Kerse Fm
Tacket Wood
Thurlestone
Preston
The Grange
East Charleton
Court Park
South Milton
Cemy
Hotel
Easton
Park Fm
Whitlocksworthy
The Croft
Auton
West Charleton
Works
Oldaway Fm
Youngcombe
Collapit Br
Wks
Horswell Ho
Sutton
Gerston Fm
The Books
Davey Park Fm
Cholwells
Woolston
Southdown
Holwell
Bagton
Gerston Point
Ham Point
Thurlestone Rock
Hotel
Court Barton
KINGSBRIDGE ESTUARY
South Huish
Wareham Point
Great Ledge
Burleigh Fm
Alston Fm
Blanksmill Br
Halwell Ho
Beacon Point
Galmpton
Ilton Castle Fm
Lincombe
Lower Barn
Woolman Point
Burton
Withymore Fm
Tosnos Point
Outer Hope
Yarde
Ilton
Shippen
Hotel
Snapes Manor
Scoble
Yeovil Rock
Inner Hope
Hope Barton
Batson
Westerncombe
Malborough
Collaton
Bolberry
Portlemore Barton
LB Sta
Whitechurch
South West Coast Path
SALCOMBE
Ford
Goodshelter
Bolberry Down
Southdown Fm
East Portlemouth
Fernyhole Point
Rew
Combe
West Cliff
Slippery Point
Hazel Tor
Holset
Castle
Catholé Cliff
Hotel
Soar
South Sands
Battery
Rickham Common
West Prawle
Lantern Rock
Splatcove Point
Rickham
Soar Mill Cove
Portlemouth Down
Sharpitor
Gara Rock
Hotel
East Soar Fm
The Bar
Ham Stone
The Bull
The Warren
Deckler's Cliff
Steeple Cove
Sharp Tor
Shag Rock
Off Cove
Starehole Bay
Pig's Nose
Mew Stone
Ham Stone
BOLT HEAD
Little Mew Stone
Ball Rock
Gammon Head
Grid North
True North
Magnetic North
Diagrammatic only
Scale 1: 50 000
2 centimetres to 1 kilometre (one grid square)
Kilometres
Miles
1 kilometre = 0·6214 mile
1 mile = 1·6093 kilometres
Extract produced by Ordnance Survey 2014.
© Crown copyright 2012. All rights reserved.
Ordnance Survey, OS, the OS Symbol and Landranger are registered trademarks of Ordnance Survey, the national mapping agency of Great Britain.
Reproduction in whole or in part by any means is prohibited without the prior written permission of Ordnance Survey. For educational use only.

[BLANK PAGE]

DO NOT WRITE ON THIS PAGE

X733/75/31

Geography
Ordnance Survey Map
Item B

THURSDAY, 21 MAY

9:00 AM – 10:45 AM

The colours used in the printing of these map extracts are indicated in the four little boxes at the top of the map extract. Each box should contain a colour; if any does not, the map is incomplete and should be returned to the Invigilator.

©

Extract No 2142/139

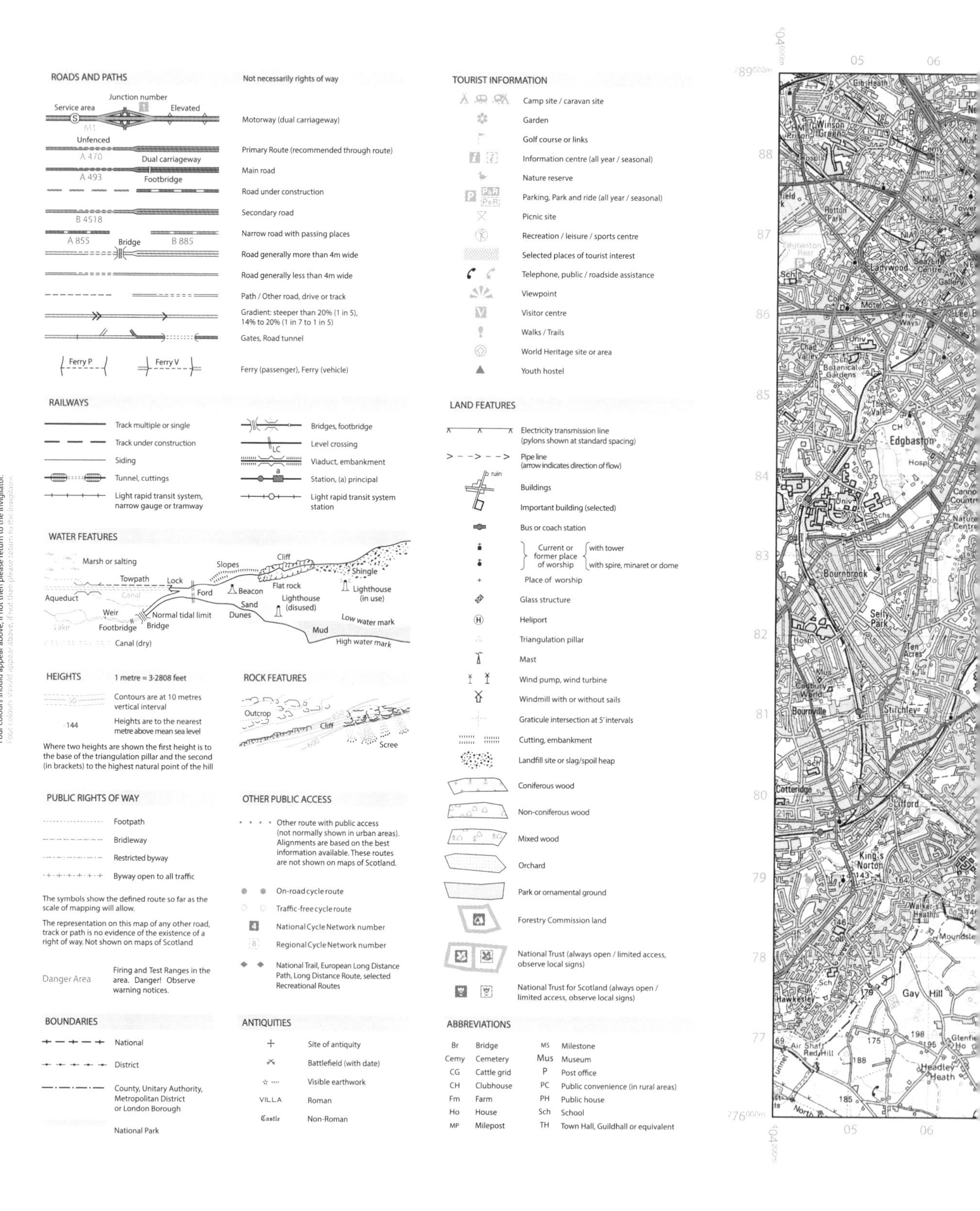

Four colours should appear above; if not then please return to the invigilator.

1:50 000 Scale Landranger Series

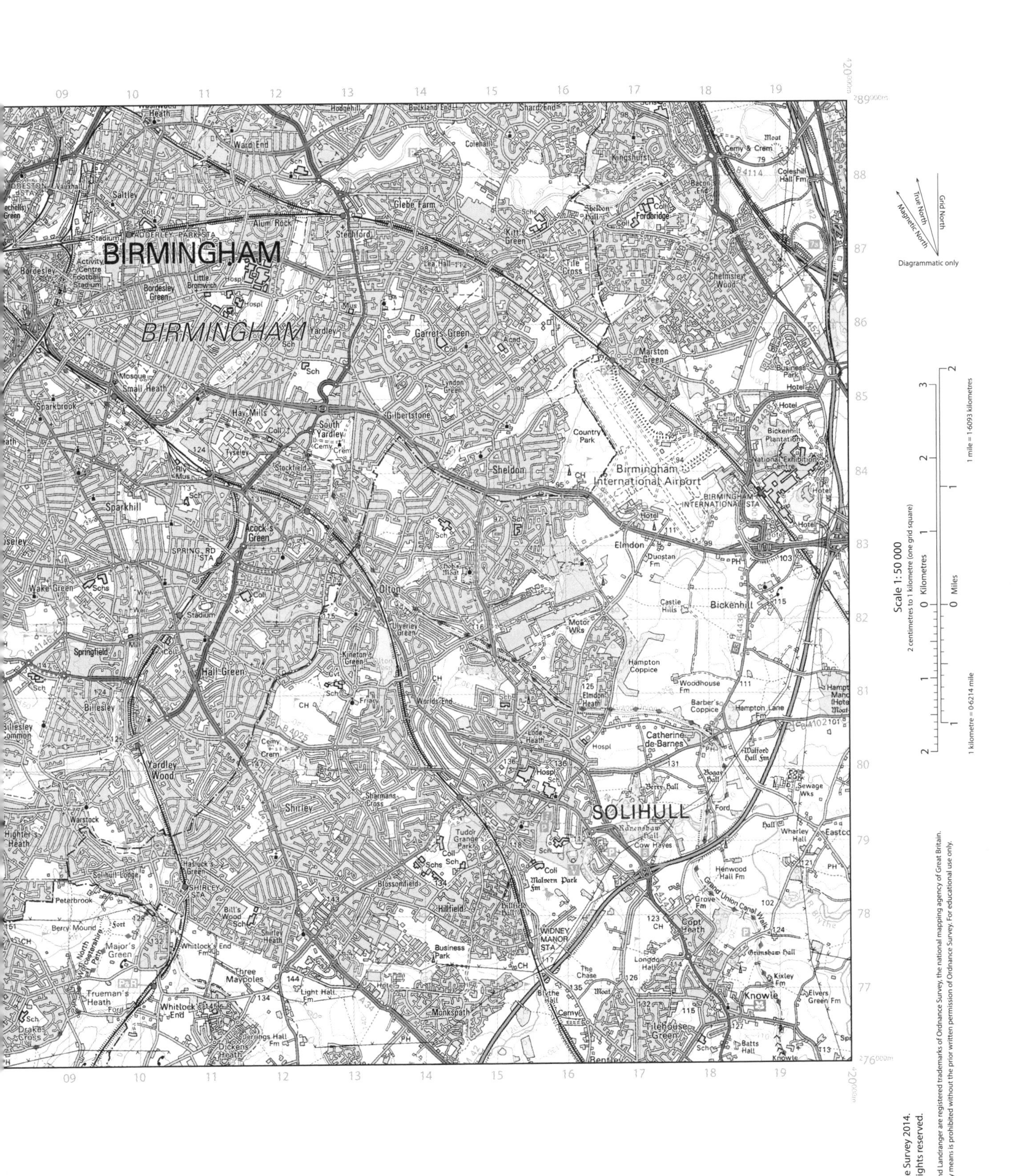

Extract produced by Ordnance Survey 2014.
© Crown copyright 2011. All rights reserved.

Ordnance Survey, OS, the OS Symbol and Landranger are registered trademarks of Ordnance Survey, the national mapping agency of Great Britain.
Reproduction in whole or in part by any means is prohibited without the prior written permission of Ordnance Survey. For educational use only.

[BLANK PAGE]

DO NOT WRITE ON THIS PAGE

NATIONAL 5

2016

National
Qualifications
2016

X733/75/11 **Geography**

FRIDAY, 6 MAY

1:00 PM – 2:45 PM

Total marks — 60

SECTION 1 — PHYSICAL ENVIRONMENTS — 20 marks

Attempt EITHER question 1 **OR** question 2. **ALSO** attempt questions 3, 4, and 5.

SECTION 2 — HUMAN ENVIRONMENTS — 20 marks

Attempt questions 6, 7 and 8

SECTION 3 — GLOBAL ISSUES — 20 marks

Attempt any **TWO** of the following

Question 9 — Climate Change

Question 10 — Impact of Human Activity on the Natural Environment

Question 11 — Environmental Hazards

Question 12 — Trade and Globalisation

Question 13 — Tourism

Question 14 — Health

Credit will always be given for appropriately labelled sketch maps and diagrams.

Write your answers clearly in the answer booklet provided. In the answer booklet you must clearly identify the question number you are attempting.

Use **blue** or **black** ink.

Before leaving the examination room you must give your answer booklet to the Invigilator; if you do not, you may lose all the marks for this paper.

©

MARKS

SECTION 1 — PHYSICAL ENVIRONMENTS — 20 marks

Attempt EITHER Question 1 OR Question 2
AND Questions 3, 4 and 5

Question 1

(a) Study the Ordnance Survey Map Extract (Item A) of the Brecon Beacons area.

Match these glaciated features with the correct grid references.

Features: **pyramidal peak**; **corrie**; **U-shaped valley**

Choose from grid references: **800217**, **927226**, **012216**, **946178**. 3

(b) **Explain** the formation of a U-shaped valley.

You may use a diagram(s) in your answer. 4

NOW ANSWER QUESTIONS 3, 4 and 5

MARKS

DO NOT ANSWER THIS QUESTION IF YOU HAVE ALREADY ANSWERED QUESTION 1

Question 2

(a) Study the Ordnance Survey Map Extract (Item A) of the Brecon Beacons area.

Match these upland limestone features with the correct grid references.

Features: **caves**; **swallow hole**; **intermittent drainage**

Choose from grid references: **891161**, **837160**, **823183**, **966146**. 3

(b) **Explain** the formation of a limestone cave/cavern.

You may use a diagram(s) in your answer. 4

NOW ANSWER QUESTIONS 3, 4 and 5

[Turn over

MARKS

Question 3

Diagram Q3: Cross-section from GR 880140 to GR 932110

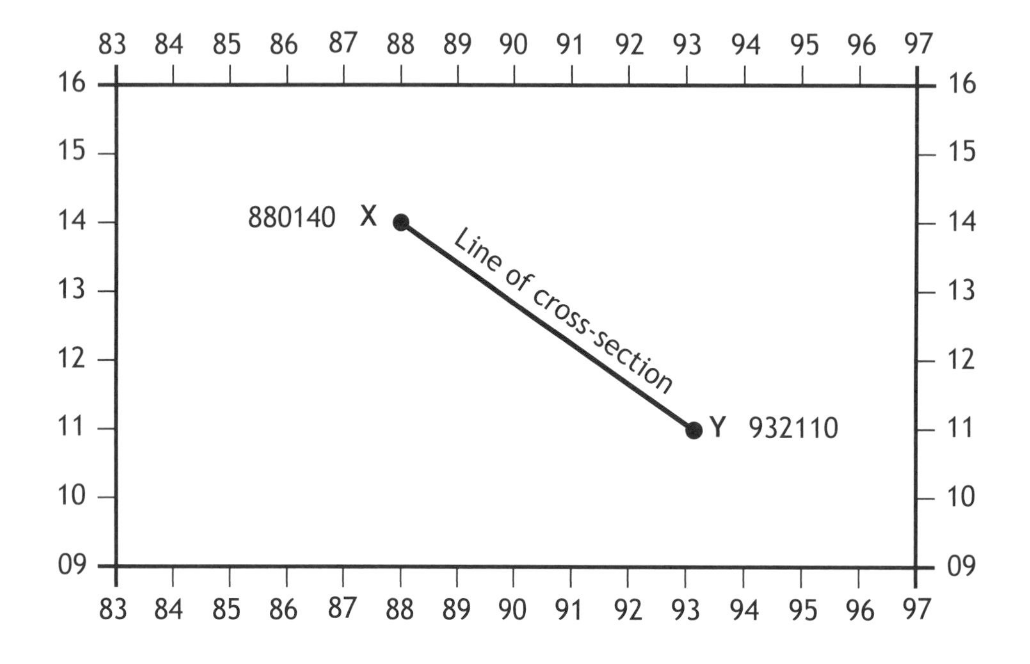

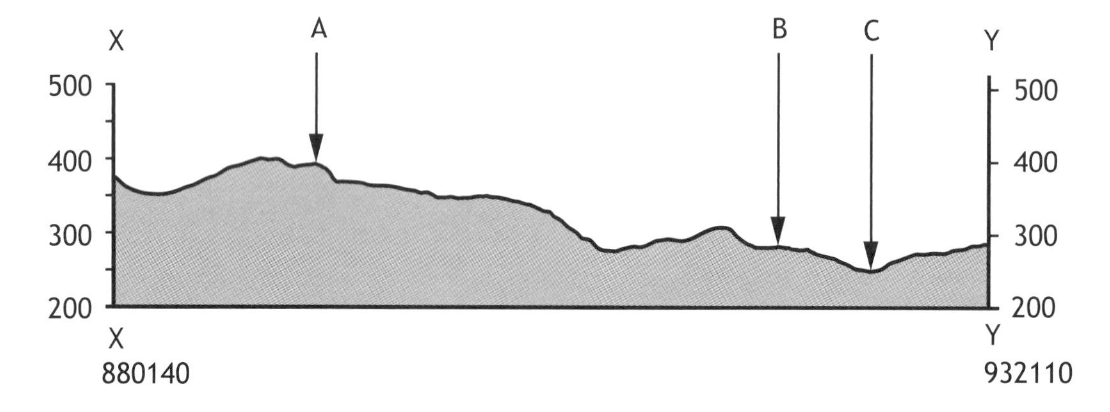

Study the Ordnance Survey Map Extract (Item A) of the Brecon Beacons area and find the cross-section X-Y shown on Diagram Q3 above.

Use the information in the OS map to match the letters A to C with the correct features, choosing from the features below.

- Afon Mellte (river)
- A4059
- Minor road
- Coniferous woodland

3

MARKS

Question 4

Diagram Q4: Quote from local resident

Study the Ordnance Survey Map Extract (Item A) of the Brecon Beacons area and Diagram Q4 above.

Choose **two** different land uses from the list shown below.

- Forestry
- Recreation and Tourism
- Farming
- Industry
- Water storage and supply
- Renewable Energy

Using map evidence, **explain** how the area shown on the map extract is suitable for your **two chosen** land uses. **5**

[Turn over

Question 5

Diagram Q5A: Actual weather conditions noon Dec 24th 2014

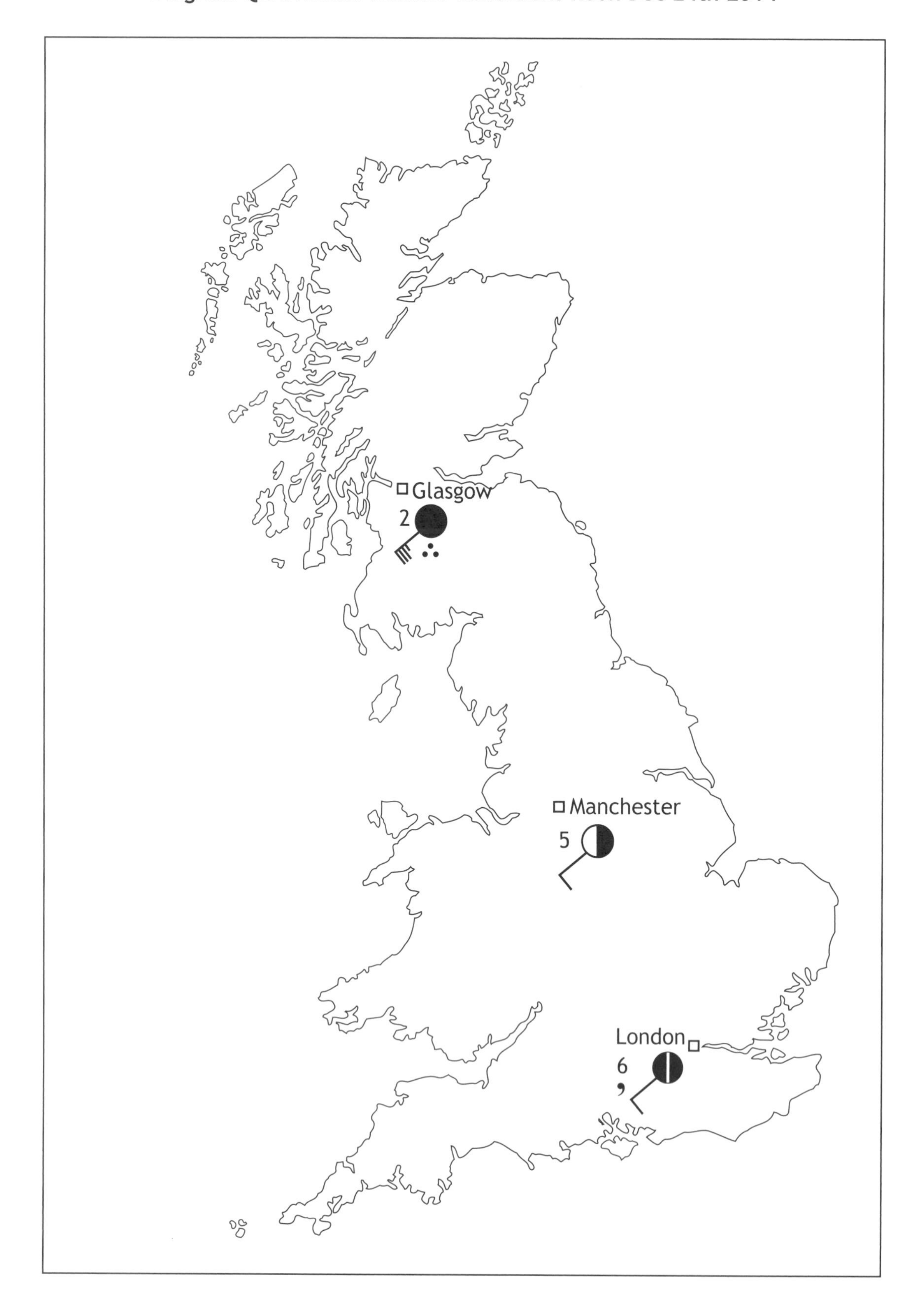

MARKS

Question 5 (continued)

Diagram 5B: Possible weather charts for noon Dec 24th 2014

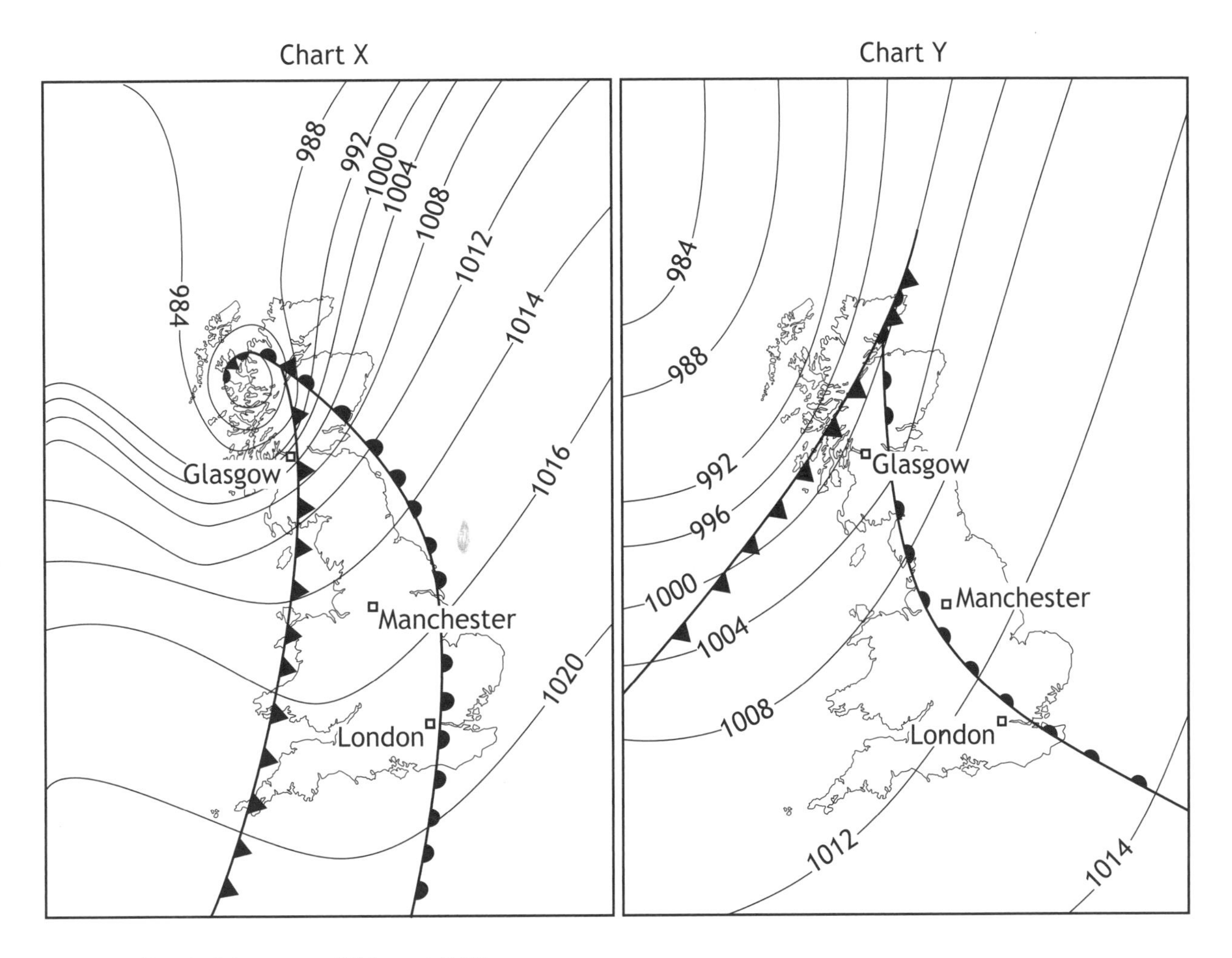

Study Diagrams Q5A and Q5B

Which weather chart, X or Y, is more likely to show the weather conditions being experienced in Britain on December 24th 2014?

Give reasons for your choice. **5**

[Turn over

MARKS

SECTION 2 — HUMAN ENVIRONMENTS — 20 marks

Attempt Questions 6, 7 and 8

Question 6

Diagram Q6A

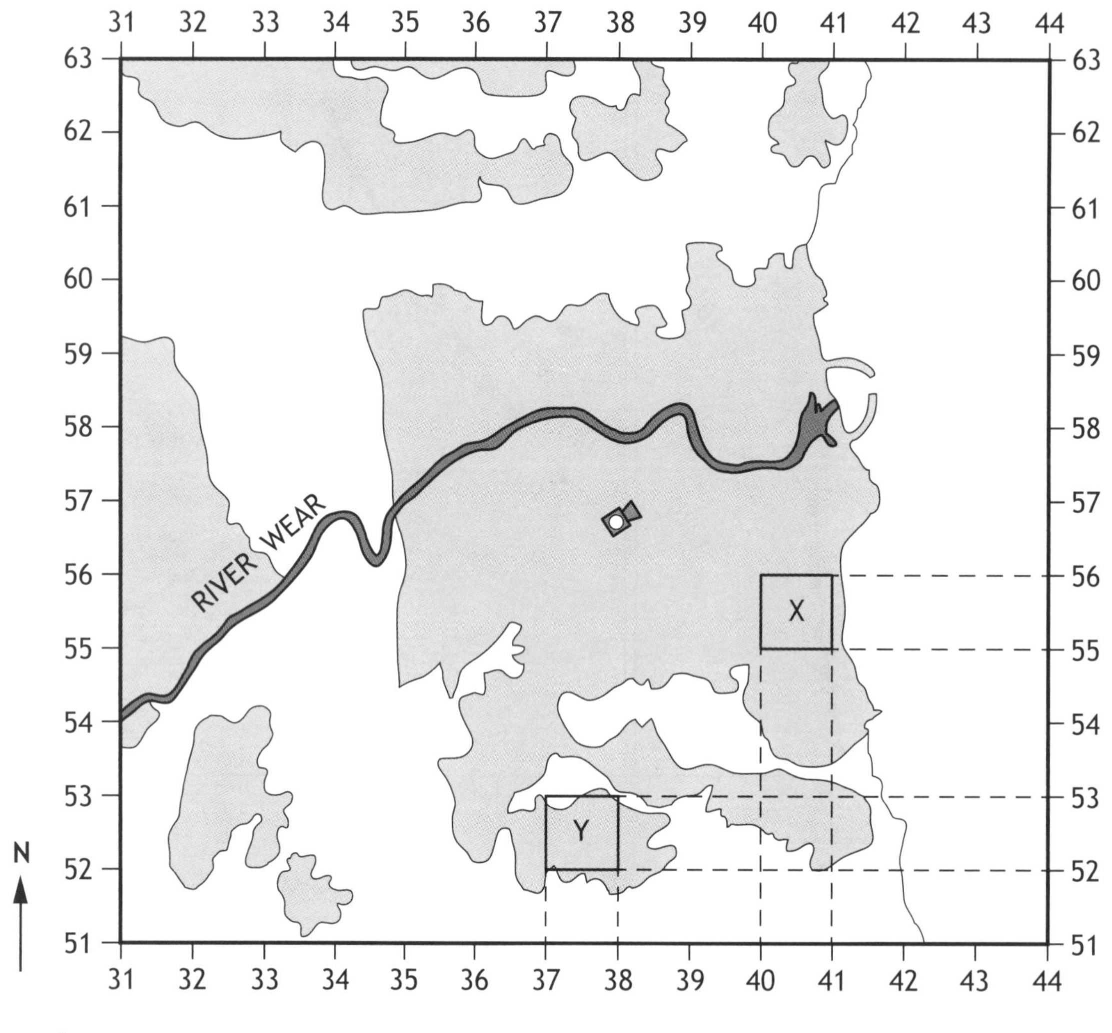

CAMERA

BUILT UP AREA

(a) Study Diagram Q6A and the Ordnance Survey Map Extract (Item B) of the Sunderland area.

A group of National 5 Geography pupils are doing a fieldwork study of urban land use for their assignment.

Use map evidence to show why they have identified Area X as the Inner City and Area Y as an area of modern suburbs. You must refer to both areas in your answer. **5**

MARKS

Question 6 (continued)

Diagram Q6B: Aerial Photograph taken from above Grid Reference 380568 looking North East

(b) Study Diagram Q6B and the Ordnance Survey Map Extract (Item B) of the Sunderland area.

Identify features A, B and C.

Choose from the list given below.

- Ayres Quay
- Docks
- Museum
- Football Stadium

3

[Turn over

MARKS

Question 7

Diagram Q7: Global Death Rates 1950 – 2030 (projected)

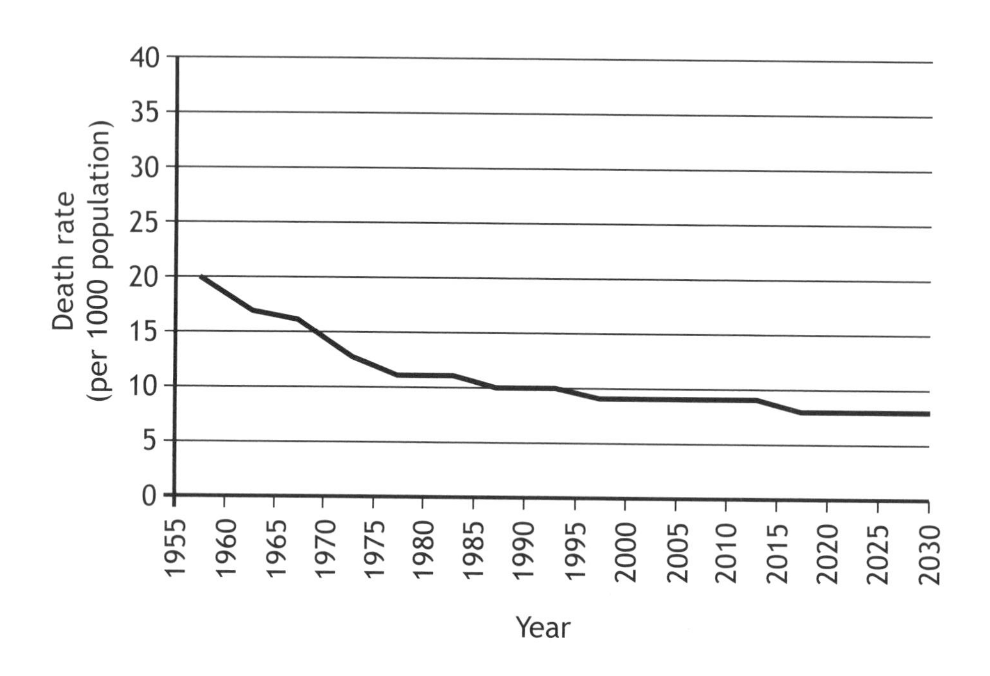

Look at Diagram Q7.

Give reasons for falling worldwide death rates. **6**

MARKS

Question 8

Diagram Q8: Modern Factors affecting Farming in the Developed World

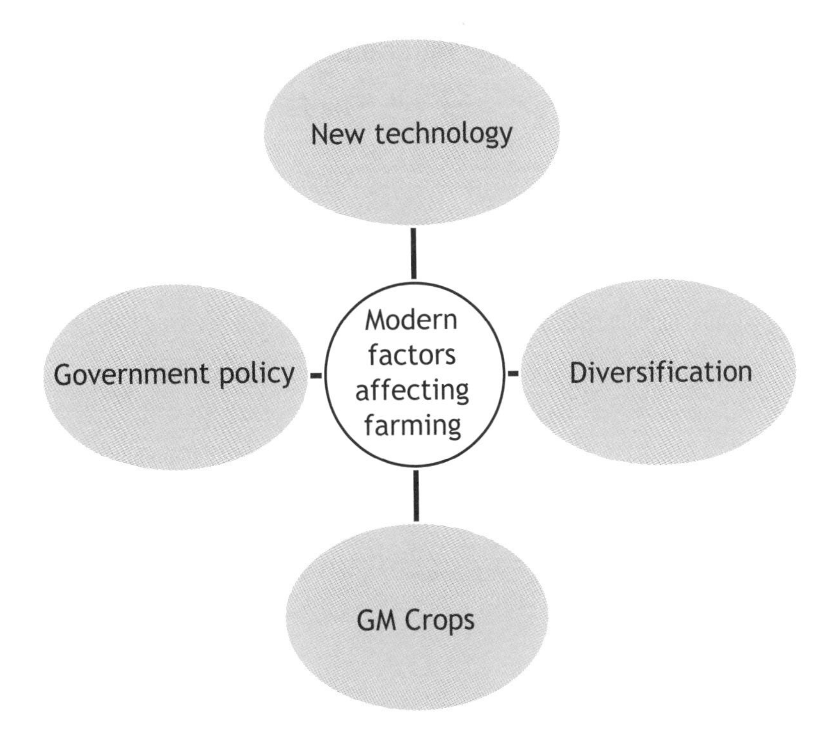

Look at Diagram Q8.

Choose **two** of the factors shown on Diagram Q8.

For the **two** you have chosen, describe in detail how recent developments in farming affect the people **and** the landscape in **developed** countries. 6

[Turn over

SECTION 3 — GLOBAL ISSUES — 20 marks

Attempt any TWO questions

MARKS

Question 9 — Climate Change

Diagram Q9A: Average global temperature changes (°C) 1884 – 2010

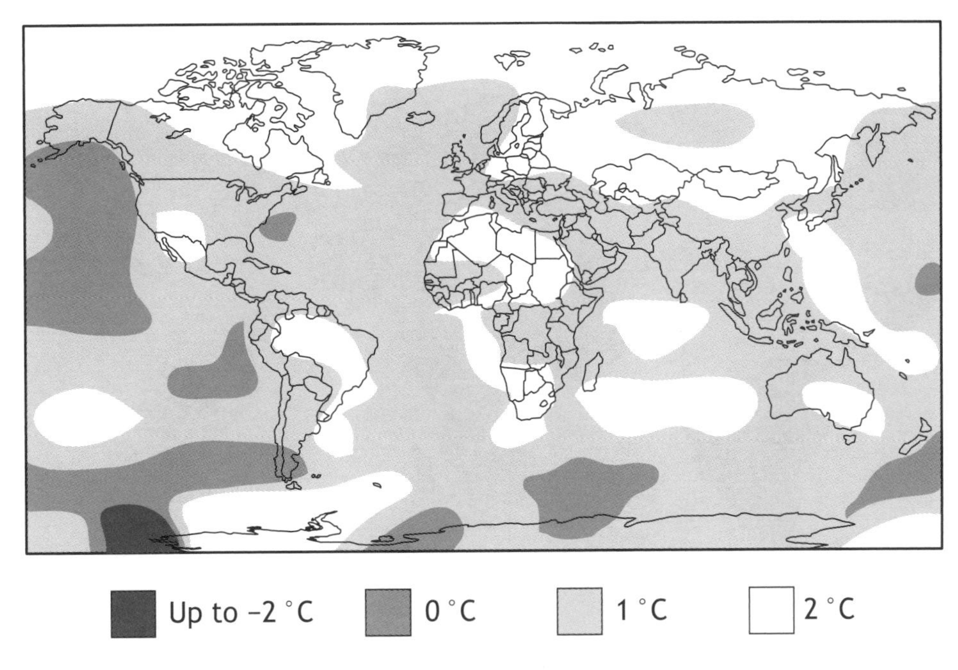

Temperature change

(a) Study Diagram Q9A.

Describe, **in detail**, the changes in average global temperatures. 4

Diagram Q9B: climate change

(b) Look at Diagram Q9B.

Describe, **in detail**, different ways climate change can be managed. 6

MARKS

Question 10 — Impact of Human Activity on the Natural Environment

Diagram Q10: Recent Deforestation Rates Worldwide

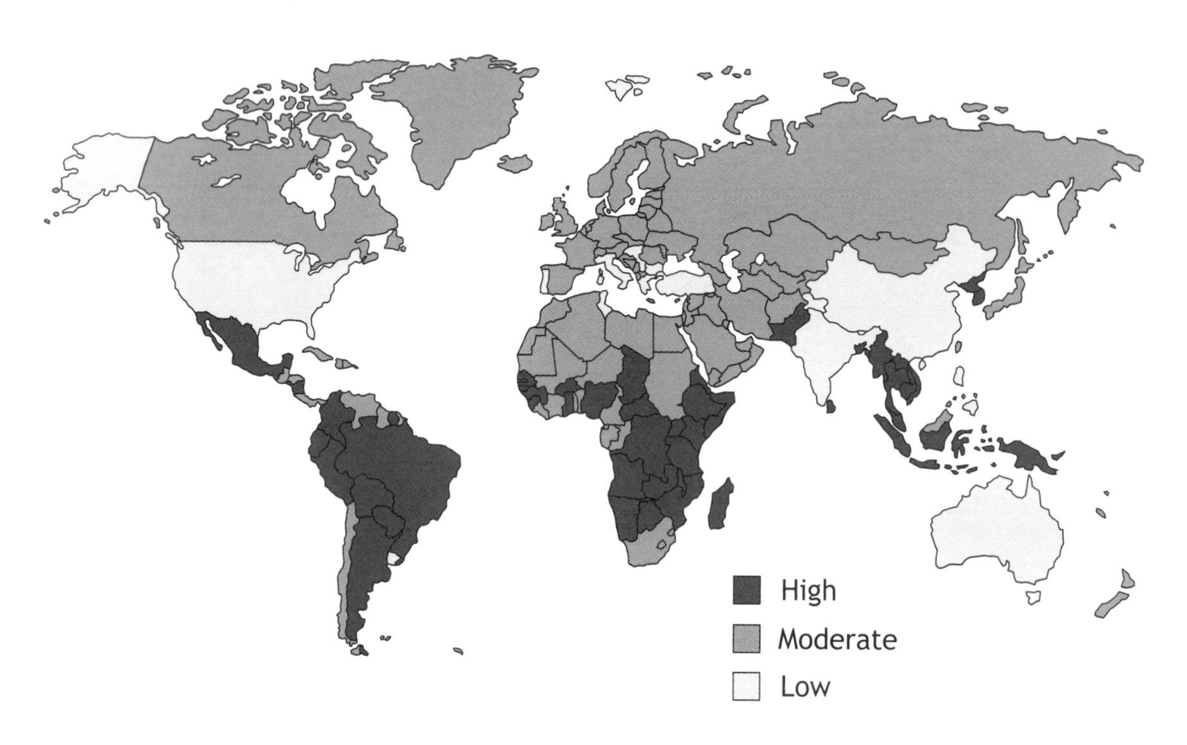

(a) Study Diagram Q10.

Describe, in detail, deforestation rates worldwide. 4

(b) **Explain** the management strategies which can be used to minimise the impact of human activity in the tundra. 6

[Turn over for next question

DO NOT WRITE ON THIS PAGE

MARKS

Question 11 — Environmental Hazards

Diagram Q11A: Number of Volcanic Eruptions per Decade 1910 – 2010

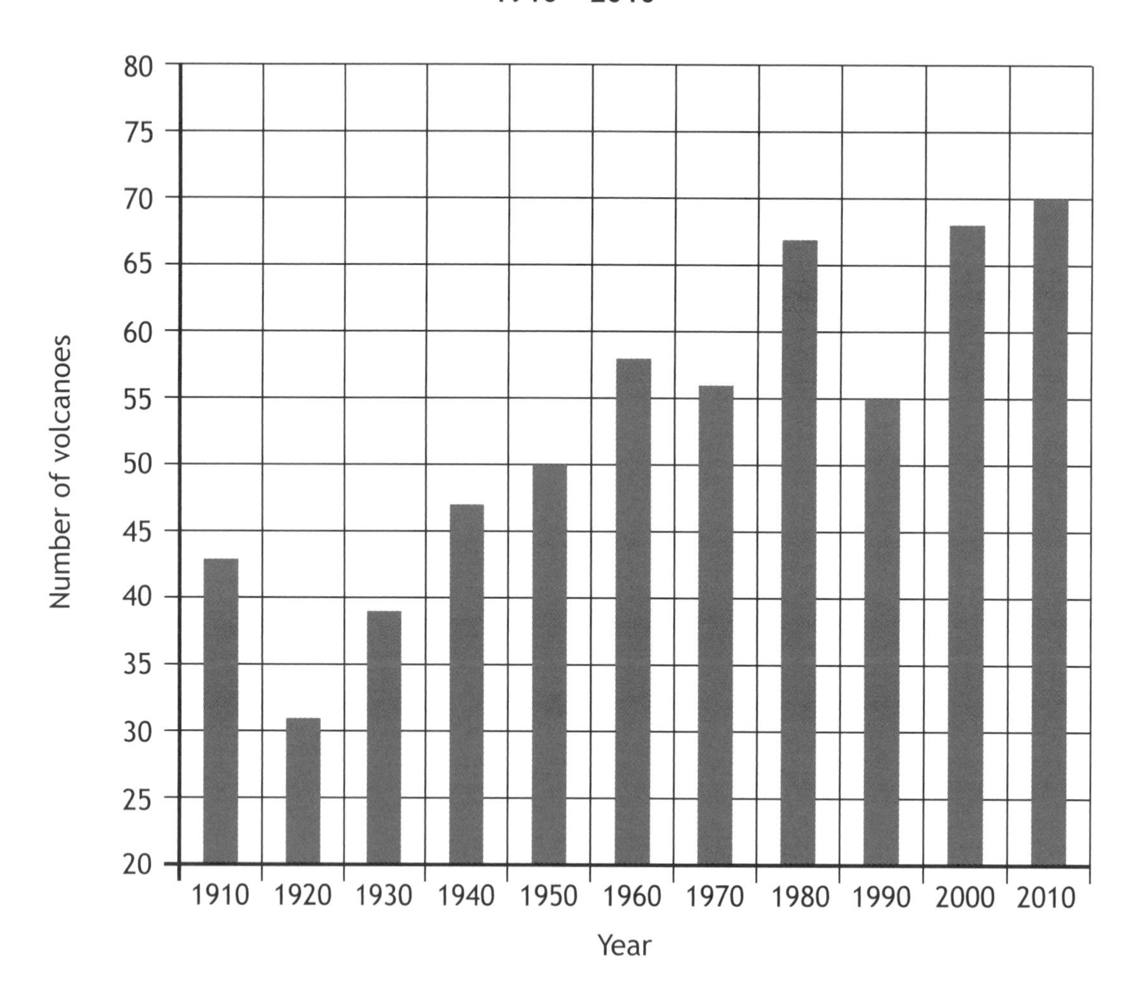

(a) Study Diagram Q11A.

Describe, **in detail**, the changes in the number of volcanic eruptions between 1910 – 2010. 4

MARKS

Question 11 (continued)

Diagram Q11B: Pico de Fogo Volcano, Cape Verde

After nearly 20 years of inactivity, the Pico de Fogo awakened with a violent eruption on the 23rd of November 2014.

(b) Look at Diagram Q11B.

For a volcanic eruption you have studied, **explain**, **in detail**, the impacts of the eruption on people **and** the landscape. 6

[Turn over

MARKS

Question 12 — Trade and Globalisation

Diagram Q12A: UK Exports 2001 – 2013

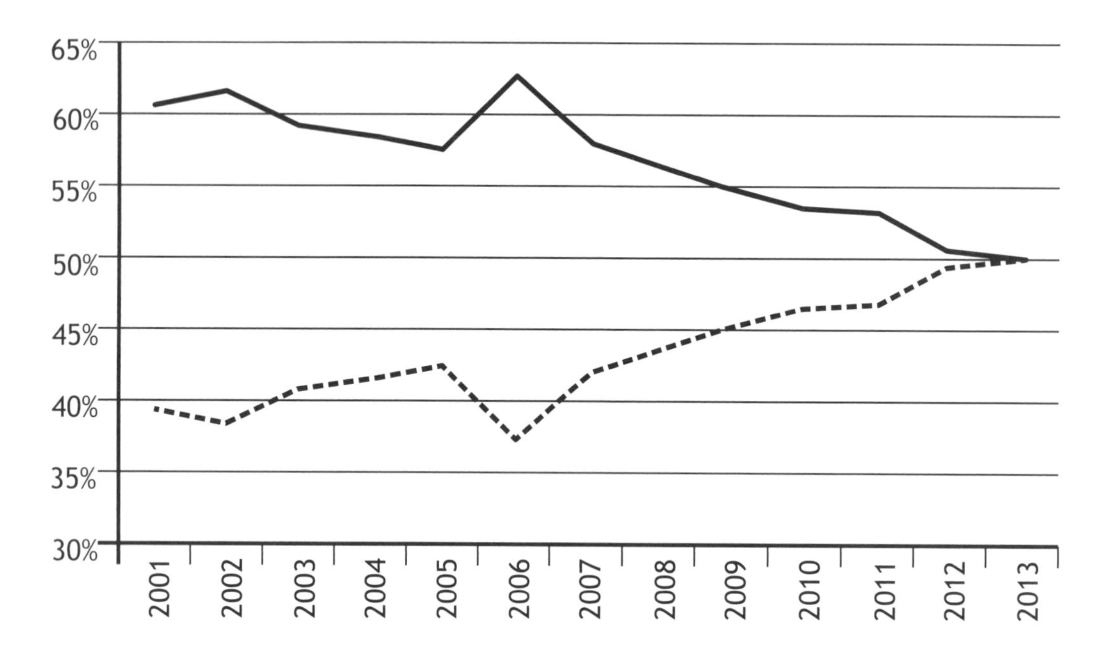

Key

% goods exports to EU27

% goods exports to Rest of the World

(a) Study Diagram Q12A.

Describe, **in detail**, the changes in UK exports between 2001 – 2013. 4

(b) **Diagram Q12B: Quote from a Government official**

"Membership of Trade Alliances such as the EU brings advantages and disadvantages".

Look at Diagram Q12B.

Explain the advantages **and** disadvantages of belonging to a trade alliance. 6

MARKS

Question 13 — Tourism

Diagram Q13: International Tourism Expenditure 2013 – 2016 (projected)

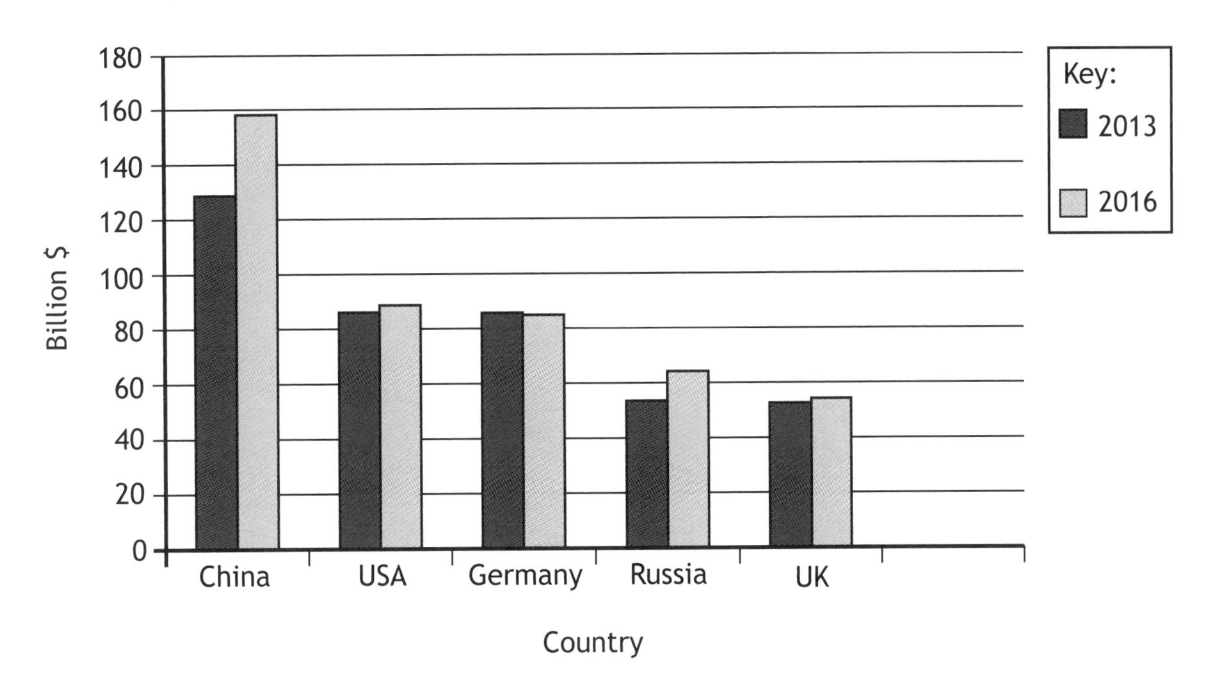

(a) Study Diagram Q13.

Describe, **in detail**, the projected changes in international tourism expenditure between 2013 and 2016. 4

(b) For a named area you have studied, **explain** the ways in which the impact of tourism can be managed. 6

[Turn over for next question

MARKS

Question Q14 — Health

Diagram Q14: Ebola Cases in Selected African Countries April-Oct 2014

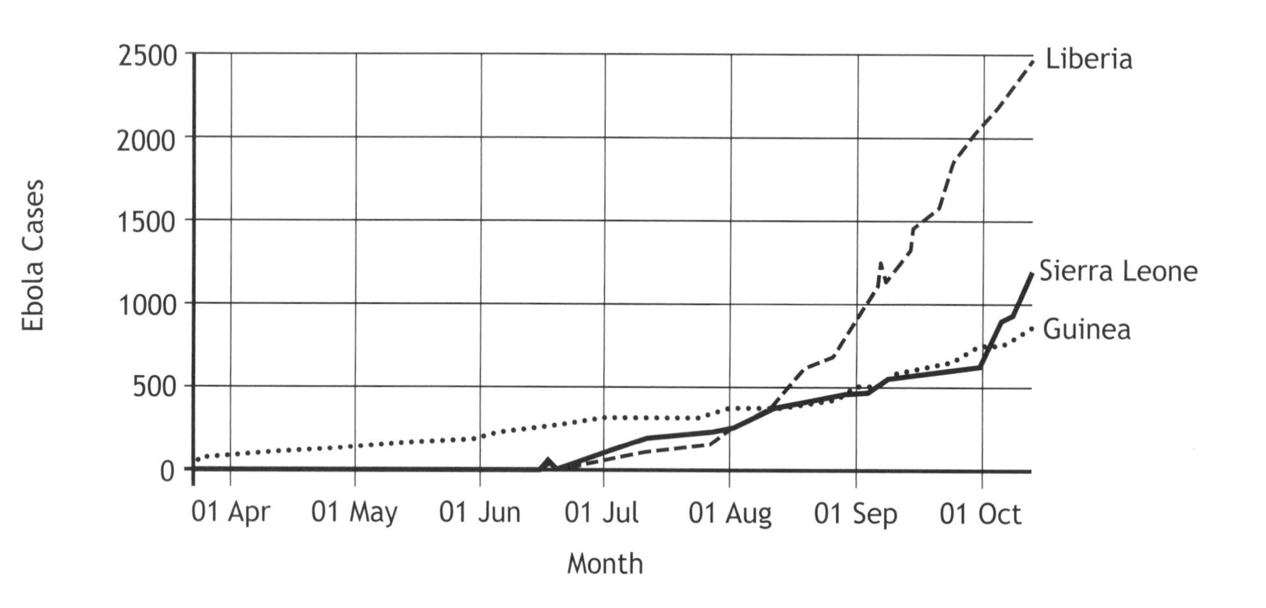

(a) Study Diagram Q14.

Describe, in detail, the changes in Ebola cases in the **three** African countries. 4

(b) **Explain** the causes of either heart disease **or** cancer **or** asthma. 6

[END OF QUESTION PAPER]

X733/75/21

Geography
Ordnance Survey Map
Item A

FRIDAY, 6 MAY

1:00 PM – 2:45 PM

The colours used in the printing of these map extracts are indicated in the four little boxes at the top of the map extract. Each box should contain a colour; if any does not, the map is incomplete and should be returned to the Invigilator.

©

1:50 000 Scale
Landranger Series

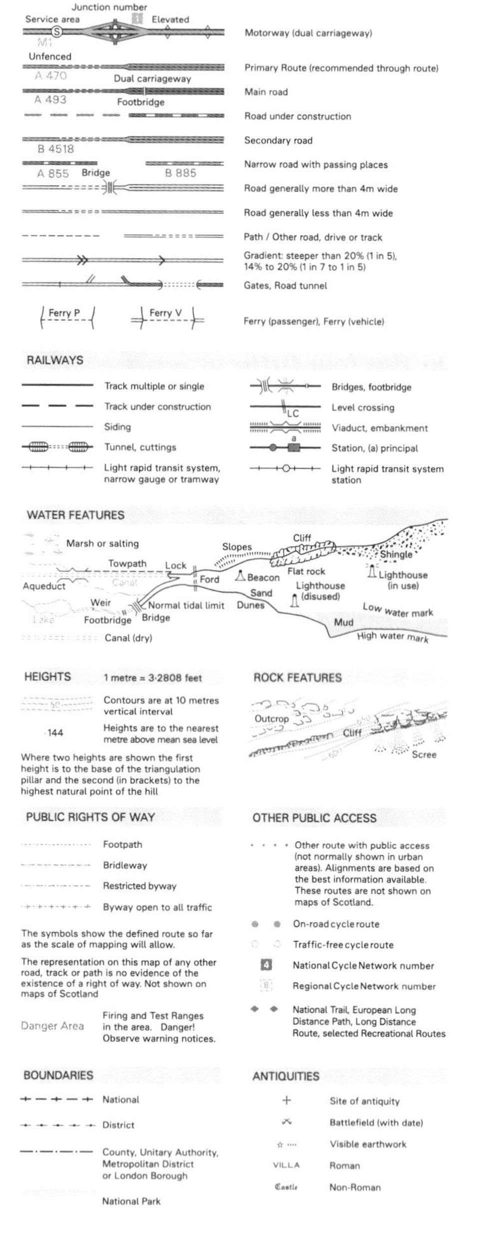

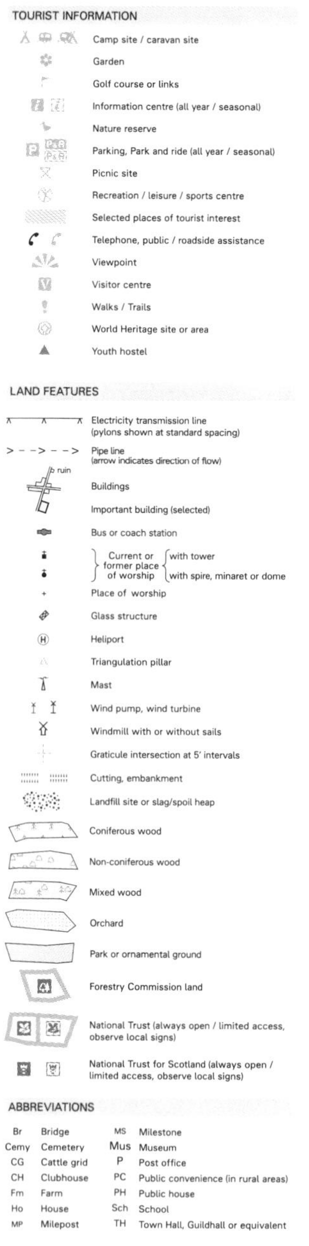

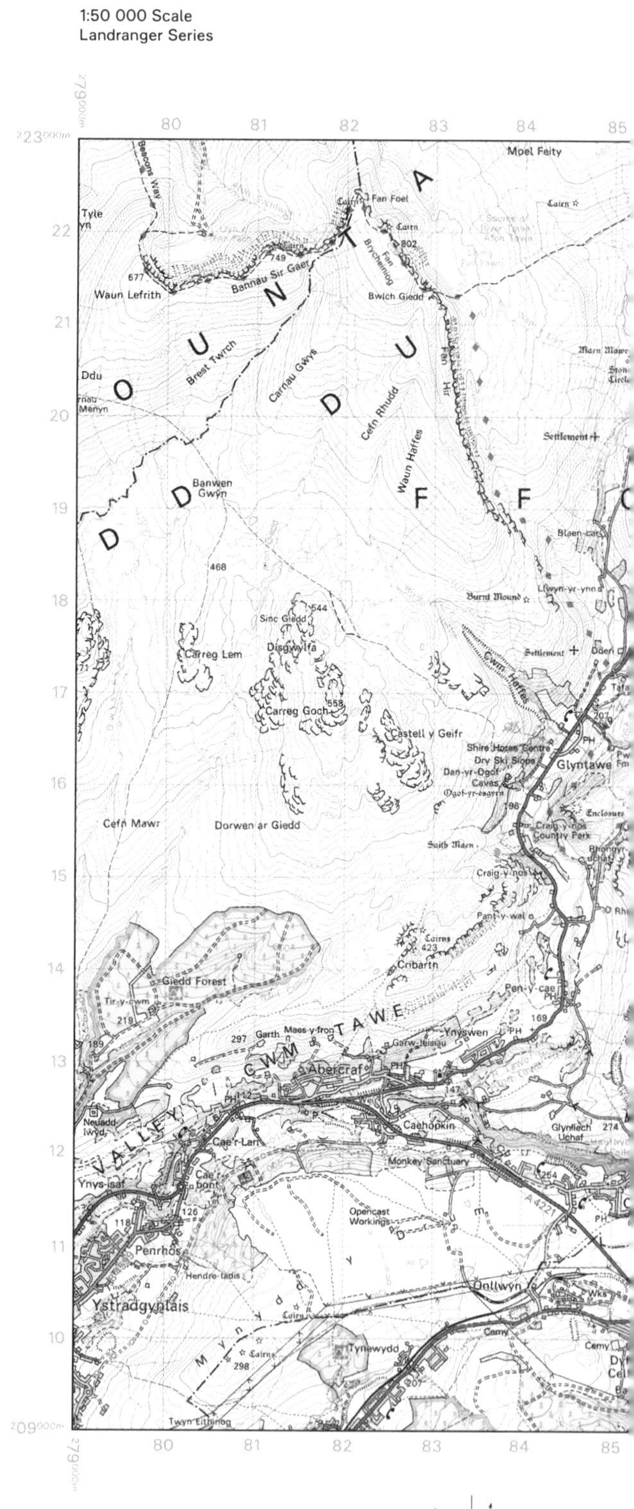

Extract produced by Ordnance Survey Limited 2015.
© Crown copyright 2013. All rights reserved.

Ordnance Survey, OS, OS logos and Landranger are registered trademarks of Ordnance Survey Limited, Britain's mapping agency.
Reproduction in whole or in part by any means is prohibited without the prior written permission of Ordnance Survey. **For educational use only.**

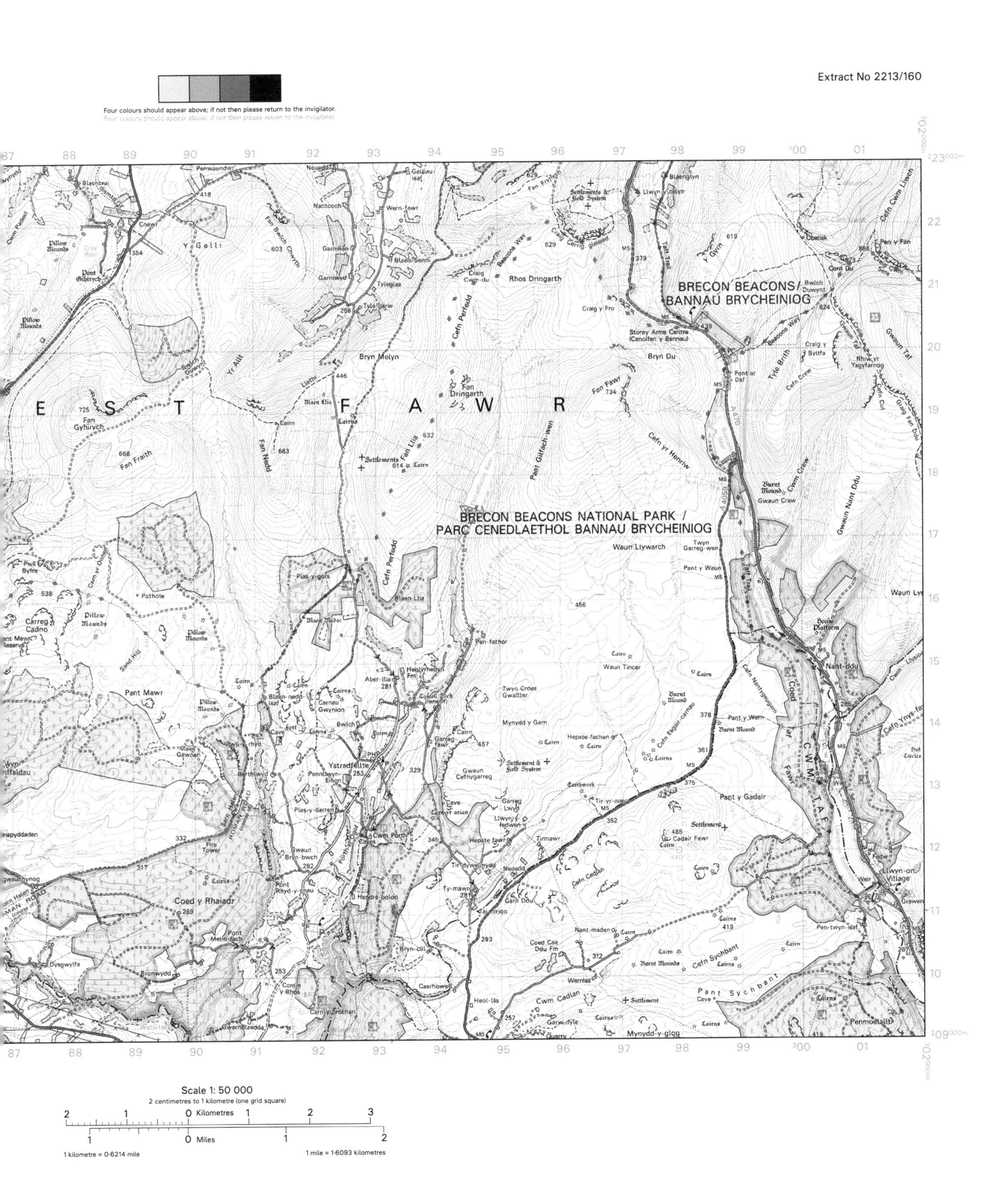
Extract No 2213/160
Four colours should appear above; if not then please return to the invigilator.
BRECON BEACONS / BANNAU BRYCHEINIOG
BRECON BEACONS NATIONAL PARK / PARC CENEDLAETHOL BANNAU BRYCHEINIOG
F A W R
E S T
Storey Arms Centre (Canolfan y Bannau)
Rhos Dringarth
Bryn Melyn
Fan Fraith
Fan Nedd
Fan Llia
Cefn Perfedd
Fan Fawr
Bryn Du
Cefn yr Henriw
Waun Llywarch
Ystradfellte
Cwm Porth
Coed y Rhaiadr
Pant Mawr
Pant y Gadair
CWM TAF
Nant-ddu
Llwyn-on Village
Cefn Sychbant
Cwm Cadlan
Penmoelallt
Mynydd-y-glog
Scale 1: 50 000
2 centimetres to 1 kilometre (one grid square)
Kilometres
Miles
1 kilometre = 0·6214 mile
1 mile = 1·6093 kilometres

[BLANK PAGE]

DO NOT WRITE ON THIS PAGE

X733/75/31

Geography
Ordnance Survey Map
Item B

FRIDAY, 6 MAY

1:00 PM – 2:45 PM

The colours used in the printing of these map extracts are indicated in the four little boxes at the top of the map extract. Each box should contain a colour; if any does not, the map is incomplete and should be returned to the Invigilator.

1:50 000 Scale
Landranger Series

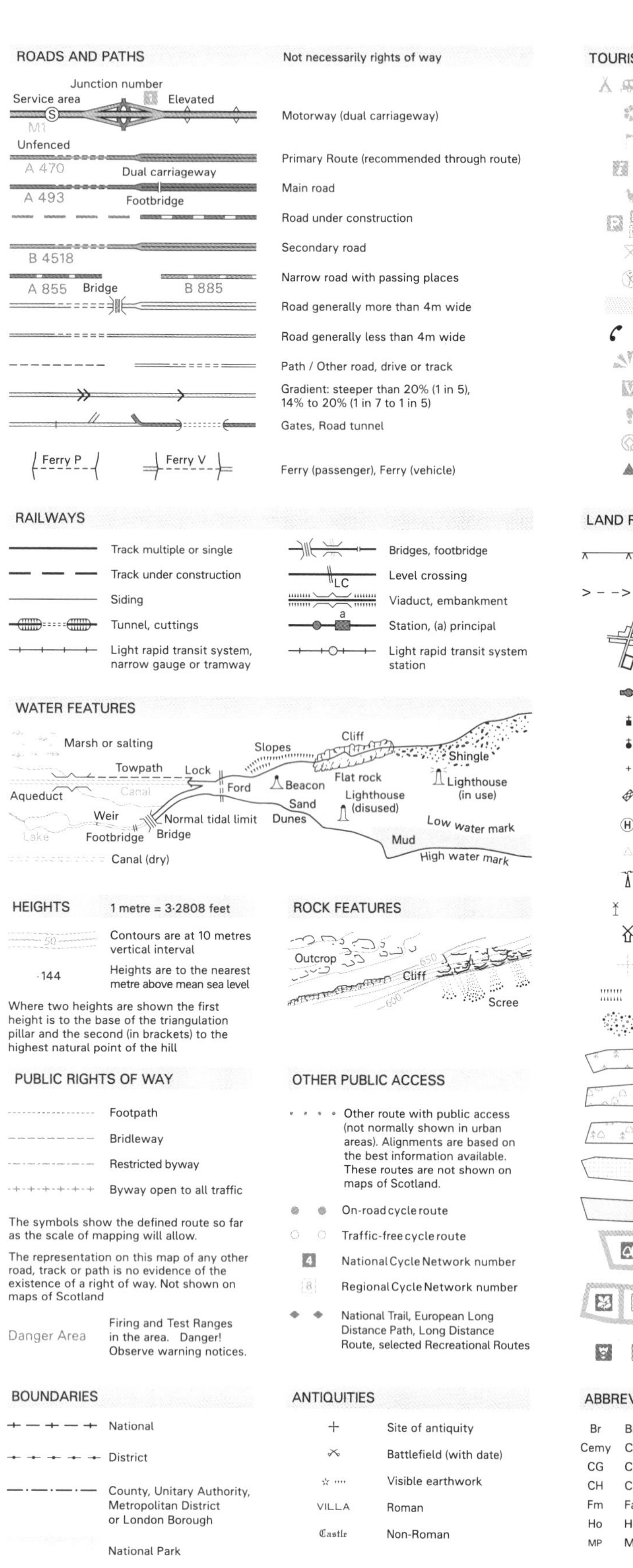

TOURIST INFORMATION

- Camp site / caravan site
- Garden
- Golf course or links
- Information centre (all year / seasonal)
- Nature reserve
- Parking, Park and ride (all year / seasonal)
- Picnic site
- Recreation / leisure / sports centre
- Selected places of tourist interest
- Telephone, public / roadside assistance
- Viewpoint
- Visitor centre
- Walks / Trails
- World Heritage site or area
- Youth hostel

LAND FEATURES

- Electricity transmission line (pylons shown at standard spacing)
- Pipe line (arrow indicates direction of flow)
- Buildings
- Important building (selected)
- Bus or coach station
- Current or former place of worship with tower / with spire, minaret or dome
- Place of worship
- Glass structure
- Heliport
- Triangulation pillar
- Mast
- Wind pump, wind turbine
- Windmill with or without sails
- Graticule intersection at 5' intervals
- Cutting, embankment
- Landfill site or slag/spoil heap
- Coniferous wood
- Non-coniferous wood
- Mixed wood
- Orchard
- Park or ornamental ground
- Forestry Commission land
- National Trust (always open / limited access, observe local signs)
- National Trust for Scotland (always open / limited access, observe local signs)

ABBREVIATIONS

Br	Bridge	MS	Milestone
Cemy	Cemetery	Mus	Museum
CG	Cattle grid	P	Post office
CH	Clubhouse	PC	Public convenience (in rural areas)
Fm	Farm	PH	Public house
Ho	House	Sch	School
MP	Milepost	TH	Town Hall, Guildhall or equivalent

Extract produced by Ordnance Survey Limited 2015.
© Crown copyright 2012. All rights reserved.

Extract No 2212/88

Four colours should appear above; if not then please return to the invigilator.
Four colours should appear above; if not then please return to the invigilator.

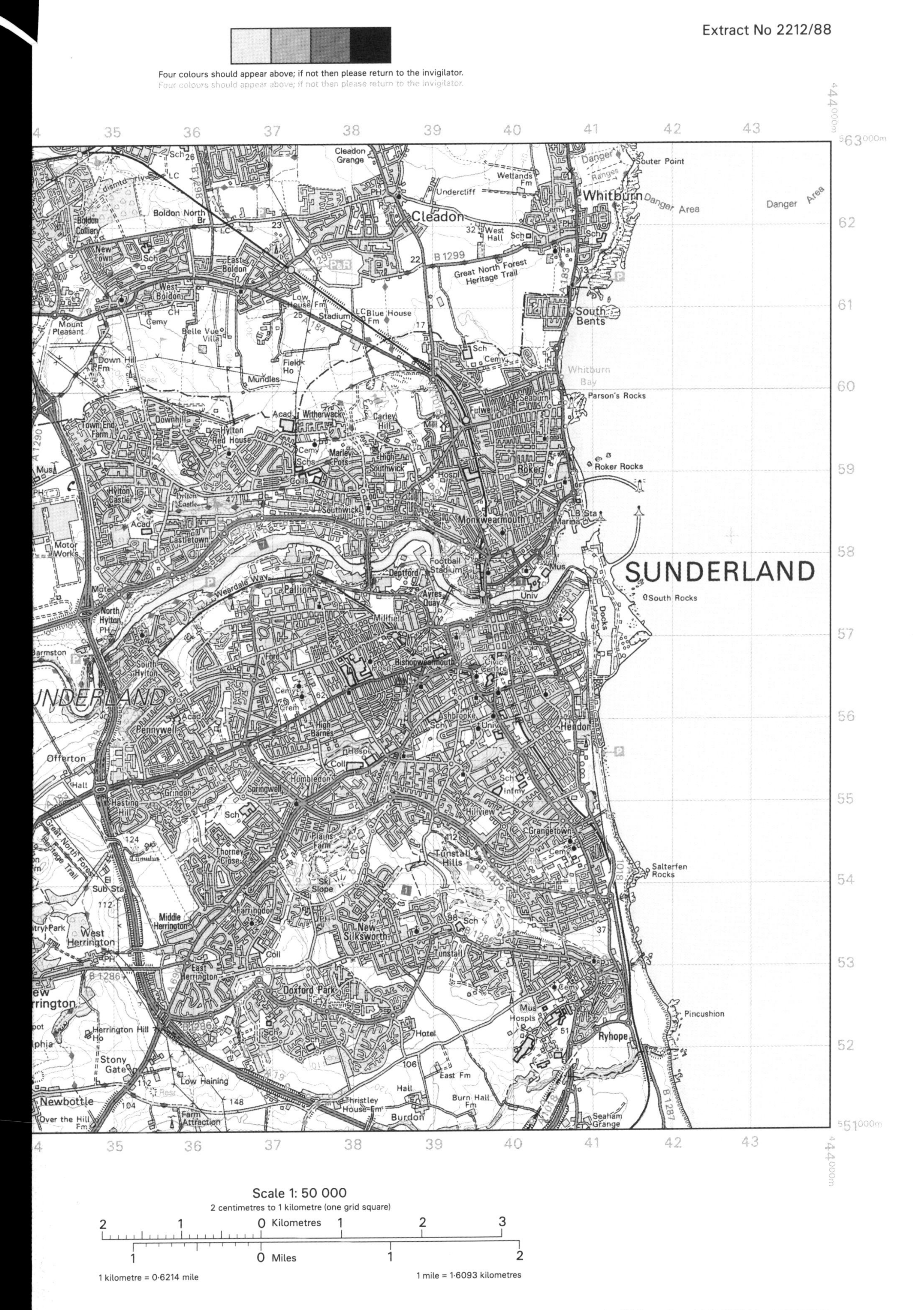

Scale 1: 50 000
2 centimetres to 1 kilometre (one grid square)

2 1 0 Kilometres 1 2 3

1 0 Miles 1 2

1 kilometre = 0·6214 mile

1 mile = 1·6093 kilometres

Ordnance Survey, OS, OS logos and Landranger are registered trademarks of Ordnance Survey Limited, Britain's mapping agency.
Reproduction in whole or in part by any means is prohibited without the prior written permission of Ordnance Survey. **For educational use only.**

[BLANK PAGE]

DO NOT WRITE ON THIS PAGE

NATIONAL 5

Answers

ANSWERS FOR

SQA NATIONAL 5 GEOGRAPHY 2016

General Marking Principles for National 5 Geography

Questions that ask candidates to *Describe* . . . (4–6 marks)

Candidates must make a number of relevant, factual points. These should be key points. The points do not need to be in any particular order. Candidates may provide a number of straightforward points or a smaller number of developed points, or a combination of these.

Up to the total mark allocation for this question:

- **One mark** should be given for each accurate relevant point.
- **Further marks** should be given for development and exemplification.

Question: Describe, in detail, the effects of two of the factors shown. (Modern factors affecting farming).

Example:

New technology has led to increased crop yields *(1 mark)*, leading to better profits for some farmers *(second mark for development)*.

Questions that ask candidates to *Explain* . . . (4–6 marks)

Candidates must make a number of points that make the process/situation plain or clear, for example by showing connections between factors or causal relationships between events or processes. These should be key reasons and may include theoretical ideas. There is no need for any prioritising of these reasons. Candidates may provide a number of straightforward reasons or a smaller number of developed reasons, or a combination of these. The use of the command word 'explain' will generally be used when candidates are required to demonstrate knowledge and understanding. However, depending on the context of the question the command words 'give reasons' may be substituted.

If candidates produce fully labelled diagrams they may be awarded up to full marks if the diagrams are sufficiently accurate and detailed.

Up to the total mark allocation for this question:

- **One mark** should be given for each accurate relevant point.
- **Further marks** should be given for developed explanations.

Question: Explain the formation of a U-shaped valley.

Example:
A glacier moves down a main valley which it erodes (*1 mark*) by plucking, where the ice freezes on to fragments of rock and pulls them away (*second mark for development*).

Questions that ask candidates to *Give reasons* . . . (4–6 marks)

Candidates must make a number of points that make the process/situation plain or clear, for example by showing connections between factors or causal relationships between events or processes. These should be key reasons and may include theoretical ideas. There is no need for any prioritising of these reasons. Candidates may provide a number of straightforward reasons or a smaller number of developed reasons, or a combination of these. The command words 'give reasons' will generally be used when candidates are required to use information from sources. However, depending on the context of the question the command word 'explain' may be substituted.

Up to the total mark allocation for this question:

- **One mark** should be given for each accurate relevant point.
- **Further marks** should be given for developed reasons.

Question: Give reasons for the differences in the weather conditions between Belfast and Stockholm.

Example:
In Stockholm it is dry, but in Belfast it is wet because Stockholm is in a ridge of high pressure whereas Belfast is in a depression (*1 mark*). Belfast is close to the warm front and therefore experiencing rain (*second mark for development*).

Questions that ask candidates to *Match* (3–4 marks)

Candidates must match two sets of variables by using their map interpretation skills.

Up to the total mark allocation for this question:

One mark should be given for each correct answer.

Question: Match the letters A to C with the correct features.

Example: A = Forestry (*1 mark*)

Questions that ask candidates to *Give map evidence* (3–4 marks)

Candidates must look for evidence on the map and make clear statements to support their answer.

Up to the total mark allocation for this question:

Question: Give map evidence to show that part of Coventry's CBD is located in grid square 3379.

Example: Many roads meet in this square (*1 mark*).

Questions that ask candidates to *Give advantages and/or disadvantages* (4–6 marks)

Candidates must select relevant advantages or disadvantages of a proposed development and show their understanding of their significance to the proposal. Answers may give briefly explained points or a smaller number of points which are developed to warrant further marks.

Up to the total mark allocation for this question:

- **One mark** should be given for each accurate relevant point.
- **Further marks** should be given for developed points.
- Marks should be awarded for accurate map evidence.

Question: Give either advantages or disadvantages of this location for a shopping centre. You must use map evidence to support your answer.

Example: There are roads and motorways close by allowing the easy delivery of goods (*1 mark*) and access for customers (*1 mark for development*), eg the A46, M6 and M69 (*1 mark*).

NATIONAL 5 GEOGRAPHY 2014

Section 1: Physical Environments

1. (a) U-shaped valley: 323143
Corrie: 326168
Arête: 309201

(b) A diagram with fully explanatory labels may gain full marks.
Snow collects in the north facing hollow of a mountain and the bottom layers turn to ice (1); the ice begins to move and the hollow is eroded (1); Rocks frozen on to the bottom of the ice scrape away at the base of the hollow (1) this is known as abrasion (1); ice plucking is when the glacier freezes on to loosened rock and pulls if free as the glacier moves (1); frost shattering may cause material to be incorporated into the ice (1); the ice melts leaving a tarn or corrie loch (1).
Or any other valid point.

2. (a) Ox-bow lake: 427099
Meander: 423107
V-shaped valley: 297207

(b) A diagram with fully explanatory labels may gain full marks.
As water flows over a hard rock band on to a softer one it erodes the soft rock faster (1); this creates a step which the water has to flow down (1); the river creates a plunge pool at the bottom of this drop (1) where the water swirls around and undercuts the hard rock (1); this erosion could be due to hydraulic action (1) where the force of the water erodes the rock (1) or due to corrasion where the river's load causes the erosion (1); eventually the hard rock collapses because there is nothing left to support it (1) and the waterfall will move back (1); over time this creates a gorge (1).
Or any other valid point.

3. One mark for first valid grid reference.
If **forestry** chosen:
There are suitable areas for forestry around Loch Ailsh (3110) (1) because the land is not too high, mostly under 300 metres (1) and there is access for vehicles from the A class road (1); there are a number of other tracks allowing lorries to take away timber (1) and the land is probably unsuitable for most other land uses as it is quite steep (1).
Or any other valid point.

If **water storage and supply** chosen:
This is an excellent area for water supply as there are a number of natural lochs such as Loch Ailsh (1) and deep narrow valleys which could be dammed (1) such as the Abhainn Gleann na Muic at 393130 (1); there is not much evidence of human activity so there are less likely to be objections to a dam (1); there are many streams and rivers (such as in 2914) indicating plentiful water supply (1).
Or any other valid point.

If **recreation/tourism** chosen:
This area would be good for recreation and tourists because there are lots of opportunities for hillwalking and climbing (1) such as on Ben More Assynt (1); there are chambered cairns eg 299103 (1) and other historic sites in the forests around Loch Ailsh (1); there are some roads for access such as the A class road/minor road (1) but mostly there are no signs of human activity and this would be an added attraction for many visitors (1) as they would appreciate the magnificent and unspoiled scenery (1).
Or any other valid point.

4. No marks awarded for description of the weather.
(a) A tropical continental air mass will bring hot dry weather in summer which could result in droughts (1); there might need to be hosepipe bans (1); grass might wither and die causing problems for livestock farmers (1); ice cream sales might rise (1) as people make the most of the sunny weather and head for the beach (1); it could be very hot and difficult to do physical work outside (1); heavy rain from thunderstorms might cause flash floods (1).
Or any other valid point.

(b) It is much windier at Ballycastle than Lerwick because the isobars are much closer together in Ireland (1); it is raining heavily at Ballycastle because it is next to the cold front whereas Lerwick is not close to any fronts (1); the wind direction is north west at Ballycastle and south east at Lerwick because the wind goes anticlockwise around the low pressure and the isobars show the approximate direction (1); the cloud cover is 8 oktas at Ballycastle because it is so close to the cold front where there is likely to be cumulonimbus cloud whereas Lerwick is not near any fronts so is less likely to have cloud (1); the temperature is warmer at Ballycastle as it is still in the warm sector whereas Lerwick is not (1); the weather is mostly worse in Ballycastle than Lerwick as it is closer to a depression (1). It is warmer in Ballycastle because the clouds trap the heat whereas it is colder in Lerwick because there are no clouds (1).
Or any other valid point.

Section 2: Human Environments

5. For full marks answer must refer to a named city.
If Glasgow chosen:
Many old buildings like Candleriggs Fruit Market are no longer needed or suited to their original purpose so are converted into houses, hotels etc (1), to make use of the valuable space (1). These converted buildings also afford the opportunity for new land users to move in (1). Many areas like Buchanan Street have become pedestrianised to make it safer for shoppers (1) and reduce the number of vehicles in the centre as well as reducing pollution (2). Many small shops have gone out of business and have been replaced with larger chain stores (1) as these land users can afford the high rents of the city centre (1). Indoor shopping malls eg Buchanan Galleries are being built to try to encourage customers back into the town centre (1). They are not affected by the weather and easily accessible to more customers (1).
Accept any other valid point.

6. Answers must refer to changes or trends.
(a) The percentage of children aged between 0 and 14 steadily decreases between 1982 and 2050 by 18% (1). The % of working age population aged between 15 and 59 increased between 1982 and 2000 by 9% (1) but is predicted to fall from 2000 to 2050 by 10% (1). The % of the population over 60 has risen constantly since 1982 by 19% (1) and the highest predicted increase of 9% between 2020 and 2050 (1).
Accept any other valid point.

(b) China used the one child policy to help reduce growth (1) the government took away benefits from families if they didn't follow this rule (1), eg increased access to education for all, plus childcare and healthcare (2). China encouraged the use of birth control methods like IUDs and sterilisation (1). Abortion is legal in China and is widely used (1). In recent years China has provided more education on birth control (1). Some countries like Indonesia introduced more free contraception (1). Some countries have used sex education (1). Some countries use tax incentives to encourage smaller families (1).
Or any other valid point.

7. Answer must refer to the differences between the cities. The population of Mumbai starts off far higher than Glasgow in 1981 (1) a difference of 7.5 m (1). Mumbai's population is rising between 1981 and 2011 whilst Glasgow's population is dropping (1). The fall in Glasgow's population levels out in 2011 whilst the rise of population of Mumbai is at its steepest (1). Mumbai's population has reached 13.6m by 2018 whilst Glasgow is at its lowest at just under 0.7m (1).
Or any other valid point.

8. Pesticides reduce disease producing better crops (1) and a surplus to trade (1). Fertilisers increase crop yields (1) this leads to better profits for some farmers (1) which can lead to an increase in their standard of living (1). Mechanisation means less strenuous work for the farmer (1) and is quicker and more efficient (1).
GM crops produce a greater yield and are disease resistant so make a greater profit for the farmer (1) they can reduce the cost to the farmer of applying pesticides (1) and reduce the risk to his health (1). The growing demand for biofuels means higher crop prices and can result in the farmer getting a higher income (1) and create employment (1).
Or any other valid point.

Section 3: Global Issues

9. (a) There is a flooding risk in South Asia (1). There is a risk of drought in Western Australia (1).
Coastal areas and low lying land are at greatest risk of flooding (1), eg Bangladesh/Indonesia (1).
Areas which lack water and are at risk of suffering drought are desert and semi-arid regions (1), eg Namib-Kalahari (1).
Crop yields decrease likely in land on edge of deserts (1), eg Sahara/Sahel regions (1).
Any other valid point/location.

(b) Maximum of 4 marks for either human or physical causes.

Physical causes:
Variations in solar energy may affect global temperature (1).
Variations in the earth's orbit around the sun may cause temperature changes (1). Sulphur dioxide gas & dust particles released in a volcanic eruption can affect amount of solar energy reaching the earth (1).
Changes in ocean currents can affect temperature in different parts of the world (1).

Human causes:
The biggest contributor is gas released into the atmosphere from cars and burning fossil fuels (1).
Cow dung and decaying landfill produced harmful gases such as methane (1) which contribute to global warming (1). Deforestation causes CO_2 level to rise because there are less trees to absorb it (1) and also burning trees increases the amount of CO_2 in the atmosphere (1). Cooling units - CFCs and HCFCs previously used as coolants in fridges, freezers and air conditioners are another cause of climate change (1).
Any other valid points.

10. (a) Answer must refer to both temperature and rainfall, otherwise maximum 3 marks.
It's very cold in the Tundra and there is not much rain (1). Tundra regions have a cold desert climate - less than 250 mm rainfall/year (1). Tundra regions tend to get a small amount of rainfall each month (1).
Highest temperature is 6 ° C in July (1) and lowest is minus 28 ° C in Jan/Feb (1).
Range of temperature in the Tundra is 34 ° C (1).
Any other valid point.

(b) Answer must refer to advantages and disadvantages. Maximum of 4 marks for either advantages or disadvantages.

eg Alaska
Advantages: The Arctic's undeveloped oil/gas/mining/forestry/fishery and other wildlife resources have the potential to provide enormous opportunity and wealth for the future (1). Oil has brought many benefits to the native people including jobs (1).
Development raises their standard of living (1), improving education and medical facilities (1).
Development of resources can help economic expansion (1).

Disadvantages: Any oil spill has serious impacts on habitat of seabirds, fish, and marine mammals (1) which could take decades to recover (1).
WWF concerned since there is no effective method for containing and cleaning up any oil spill in ice conditions (1).
The Gwich'in Indians are concerned about the threat to their culture and way of life (1). Polar bears are also threatened with extinction as they are forced out of their natural habitat (1). Developments such as oil pipe lines affect migration routes of caribou (1).
Any other valid point.

11. (a) Most volcanic activity is found on plate boundaries (1). Many volcanoes are located on the west coast of South America (1) where the Nazca plate meets the South American plate and is forced underneath it (1).
Many volcanoes are found along the mountain ranges of the Andes in South America and the Rockies in North America (2).
Accept any other valid point.

(b) For full marks both the people and the landscape must be mentioned.
If **Japanese Earthquake** chosen:
The earthquake caused a tsunami which flooded the land (1). Cars, ships and buildings were swept away by the wall of water (1). Nuclear reactor plant damaged (1). Thousands of people living near the Fukushima nuclear power plant had to evacuate (1).
A 10m wave struck Sendai, deluging farmland (1) and sweeping cars across the airport's runway (1). Fires

broke out in the centre of the city (1). A ship carrying 100 people was swept away off the coast (1). A dam burst in north-eastern Fukushima prefecture, sweeping away homes (1). About four million homes in and around Tokyo suffered power cuts (1). Thousands of people killed (1) or made homeless due to buildings collapsing (1).
Or any other valid point.

12. (a) Europe dominated World Trade Exports with around 43% in 2005 (1).
This had dropped to around 38% in 2010 (1). Europe still remains the largest exporter in 2010 (1).
Asia had the second largest regional share of World Trade with around 27% in 2005 (1), growing to around 31% in 2010 (1).
Africa's share is low, around 3%, (1) but has grown by about 1% (1).
North America's share has dropped from just under 15% in 2005 to around 14% in 2010 (1).
Or Any other valid point.

(b) Farmers are paid a fair wage for their work (1) and safer working conditions are promoted (1). The 'middle man' is removed, (1) meaning more money is paid to the local people (1).
Money from fair trade can be used to improve services in local communities (1) such as schools and clinics (1) which improves standard of living (1).
Or any other valid point.

13. (a) The USA has six out of the ten most popular tourist attractions in the world including Niagra Falls and Disneyland (1). The most visited tourist attraction is Times Square in the USA with 35 million visitors per year (1). Washington D.C. is the second most popular tourist destination with 25 million visitors (1). Trafalgar Square is the most popular tourist area in Europe (1). Notre Dame and Disneyland in Paris are the most visited attractions in France with 12 million and 10.6 million visitors a year (2). Disneyland, Tokyo is the most visited attraction in Asia (1). Four out of the top ten most popular tourist destinations are Disneyland/Disneyworld parks located on 3 different continents (1).
Or any other valid point.

(b) Maximum of 4 marks for either effects on people or environment.

People (positive): Local people are employed to build tourist facilities eg hotels (1) and work in restaurants and souvenir shops (1). Employment opportunities allow locals to learn new skills (1) eg obtain a foreign language (1) and earn money to improve their standard of living (1). Services are improved and locals can benefit by using tourist facilities such as restaurants and water parks (1). Better employment opportunities increases the local Governments' revenue as wages are taxed (1) so they can invest in schools, healthcare and other social services (1). Locals can experience foreign languages and different cultures (1) and can also benefit from improvements in infrastructure eg roads and airports (1).

People (negative): Tourist-related jobs are usually seasonal therefore some people may not have an income for several months (1) eg at beach and ski resorts (1). Large numbers of tourists can increase noise pollution and upset the peace and quiet (1). Local people may not be able to afford tourist facilities as visitor prices are often higher than local rates (1). Tourists can conflict with local people due to different cultures and beliefs (1). There is additional sewage from visitors which increases the risk of diseases like typhoid and hepatitis (2).

Environment (positive): The appearance of some areas can be improved by modern tourist facilities (1). Some tourists are environmentally conscious and can have a positive impact on the landscape by donating money to local projects which help to protect local wildlife (1), eg nature reserves (1). Tourist beaches are cleaned up to ensure they are safe enough for people to use (1) through initiatives like Blue Flag (1). Seas become less polluted as more sewage treatment plants are built to improve water quality (1).

Environment (negative): Land is lost from traditional uses such as farming and replaced by tourist developments (1). Traditional landscapes/villages are spoiled by large tourist complexes (1). Air travel increases carbon dioxide emissions and contributes to global warming (1). Traffic congestion on local roads increases air and noise pollution (1). Tourist facilities such as large high-rise hotels and waterparks spoil the look of the natural environment (1). Litter causes visual pollution (1). Increased sewage from tourists can cause water pollution (1). Polluted water damages aquatic life and their habitats (1).
Or any other valid point.

14. (a) Maximum of 2 marks if no reference to figures.
More children under the age of 5 die in developing countries (1). African countries have the most number of child deaths under the age of 5 (1). Many African countries have between 100 and 199 child deaths under the age of 5 (1), eg Sudan (1). There are mostly between 10 and 49 child deaths per 1,000 in Asia (1), eg Russia and China (1). There are typically between 10 and 49 child deaths under the age of 5 in South America (1) except Bolivia (1) which has between 50 and 99 (1). The continents of North America and Europe have the least number of child deaths (1). There are less than 10 child deaths under the age of 5 in many developed countries (1), eg UK (1).
Or any other valid point

(b) For full marks reference must be made to developed and developing countries.
Health Education programmes have been introduced to limit the spread of AIDS in Developing and Developed countries (1). ARV drugs are also more freely available (1). Condoms are available for free (1) and TV and Radio advertising has been used to get the message across (1). Agencies such as the World Bank have made funding available to Developing Countries to tackle the disease (1). In Developed Countries, needle exchanges (1) and drug therapy programmes (1) have been introduced.

NATIONAL 5 GEOGRAPHY 2015

Section 1: Physical Environments

1. (a) Headland - 766356
Cliff - 690382
Bay - 674398

(b) **Stack**
Waves attack a line of weakness in the headland (1). Types of erosion include hydraulic action, corrosion and corrasion (1). Continuous erosion will open up the crack and it will develop into a sea cave (1). Further erosion of the cave, often on opposite sides of the headland, will form an arch (1). The roof of the arch is attacked by the waves until it eventually collapses (1). This leaves behind a free standing piece of rock called a stack which is separate from the headland (1).
Or any other valid point.

Bay
Bays are formed due to differential erosion (1) where rocks along the coastline are formed in alternating bands of different rock types (1) eg sandstone and clay (1) and which meet the coast at right angles (1). Clay is a softer rock than sandstone so it is eroded more quickly (1). The waves erode the softer rock through hydraulic action, corrasion and corrosion (1) to form sheltered bays (1) which may have beaches (1). The harder sandstone areas are more resistant to erosion and jut out into the sea to form exposed headlands (1).
Or any other valid point.

2. (a) Levée - 684466
Meander - 708473
V-Shaped Valley - 713410

(b) **Meander**
In the middle/lower course, a river flows downhill causing lateral erosion (1). The river contains areas of deep water and areas of shallow water, this results in areas of slower and faster water movement and this causes the current to swing from side to side (2). The river flows faster on the outer bank and erodes it (1). This forms a river cliff (1). The river flows more slowly on the inner bank and deposits some of its load (1). This forms a river beach/slip-off slope (1). Continuous erosion on the outer bank and deposition on the inner bank forms a meander in the river (1).
Or any other valid point.

V-shaped valley
In the upper course, a river flows downhill eroding the landscape vertically (1). The river erodes a deep notch into the landscape using hydraulic action, corrasion and corrosion (1). As the river erodes downwards the sides of the valley are exposed to freeze-thaw weathering which loosens the rocks and steepens the valley sides (2). The rocks which have fallen into the river aid the process of corrasion which leads to further erosion (1). The river transports the rocks downstream and the channel becomes wider and deeper creating a V-shaped valley between interlocking spurs (2).
Or any other valid point.

3. Answers will vary depending upon the land uses chosen.

For farming: Reads Farm (1) (at grid reference 728489) is an example of a hill sheep farm as the land is steep (1). As the land is higher up, the weather will be harsh and sheep can survive these conditions, especially in winter (1). The land is too steep for farm machinery to operate (1). The soil will be too thin for crops to be grown (1).

For tourism and recreation: The South West Coastal Path follows the top of the cliffs and allows tourists to enjoy a view of the coastal scenery (1) eg 727367 (1). There is a nature reserve for people who want to observe wildlife at 747405 (1). There is a golf course for golf enthusiasts at 668428 (1). There are various camp/caravan sites for people to stay whilst visiting the various attractions in the area (1).

4. Answers will vary depending upon the land uses chosen.

Problems between tourists and farmers:
In the Cairngorms, tourists can disrupt farming activities as walkers leave gates open, allowing animals to escape (1). Tourists' dogs can worry sheep if let off their lead (1). Stone walls are damaged by people climbing over them instead of using gates/stiles (1). Noisy tourists can disturb sheep especially during breeding season (1). Farmers may restrict walkers access at certain times eg lambing season (1). Farm vehicles can slow up tourist traffic on roads (1) and parked cars on narrow country roads can restrict the movement of large farm vehicles (1).

Problems between industry and tourists:
Tourists want to see the beautiful and unusual scenery of the Yorkshire Dales but quarries spoil the natural beauty of the landscape (1). Lorries used to remove the stone endanger wildlife and put visitors off returning to the area (1). This threatens local tourist-related jobs eg in local restaurants (1). The large lorries needed to remove the quarried stone cause air pollution which spoils the atmosphere for tourists (1). Lorries cause traffic congestion on narrow country roads which slows traffic and delays drivers (1). The peace and quiet for visitors is disturbed by the blasting of rock (1). Some wildlife habitats may also be disturbed by the removal of rock (1).
Or any other valid point.

5. South-East England is usually warmer because it is closer to the Equator (1). This is due to intense heating from the sun (1) because sun rays are more concentrated (1). Places in Northern Scotland eg Wick, are colder because they are closer to the North Pole(1). This is due to a lack of insolation from the sun as the rays are less concentrated (1) and reflection of heat by the snow and ice (1). Places located on flat low-lying land are warmer eg Central Scotland, because temperatures increase as altitude decreases and places higher up ie mountainous regions are colder (1) because temperature decreases by 1oC for every one hundred metres in height (1). Places which are south facing are warmer because they get more sun (1) and places which are north facing are colder because they experience cold northerly winds (1). Western coastal areas are warmer because of a warm ocean current (1) (The North Atlantic Drift) and due to the prevailing South-Westerly winds that are warmed as they pass across this warm ocean current (1). In summer, places closer to the sea are cooler and in winter they are warmer because the sea heats up slowly in summer and cools slowly in winter (2).
Or any other valid point.

Section 2: Human Environments

6. (a) Main roads lead into this square (1) there is a bus station (1) and two railway stations (1) tourist information centre (1) several churches (1) museum (1). Or any other valid point

(b) The land is flat so easy to build on (1) there is space available for expansion (1) eg expansion of the motor works at 163823 (1). There are good transport links like the M42 allowing people and products access to and from the area (1). A rail link with Birmingham International Rail Station gives easy access to the airport (1). There are many road junctions and intersections connecting the area to other areas and less traffic congestion as it is away from Birmingham city centre (2). The land is on the edge of Birmingham so will be cheaper encouraging housing estates like Sheldon to be built (1). The cheaper land allows the houses to be bigger with cul-de-sacs, gardens etc.(1). The houses can provide a source of labour for the airport, motor works and the business park (1).
Or any other valid point

7. Contraception and family planning is widely available (1). Later marriages are more common which results in fewer children (1). People no longer choose to have lots of children as improved medical care and advances in medicine (1) have resulted in most children surviving at birth (1). Developed countries have the money to invest in medical care which reduces the infant mortality rate thus causing the birth rate to fall (1). Children are expensive so the greater number of children the bigger the financial burden (1). Women want careers so put off having children to a later age (1) or limit the size of their families to give them a reasonable standard of living (1). Sex education in schools helps to lower birth rates (1).
Or any other valid point.

8. For example in Rocinha (Rio), the former wooden shacks have been upgraded to permanent dwellings with some modern services (1). Residents constantly improve their homes through a process of 'self-help' (1) where the residents are provided with materials like bricks (1). Some prefabricated houses have been built by the Brazilian government (1) with basic facilities like toilets, electricity and running water (1). The residents have been given the legal rights to the land (1), roads have been built into/ or improved in the favela (1) allowing services like rubbish collections to take place (1), there are now a few health clinics and schools provided (1).
Or any other valid point.

Section 3: Global Issues

9. (a) The overall trend is that the amount of Arctic Sea ice has decreased between 1979 and 2013 (1) from (around) 7 million square kilometres to (about) 5 million square kilometres (1). There has been a fluctuation in the extent of sea ice in certain years (1) eg in 2013, the amount of sea ice increased from 3.75 million square kilometres in 2012 to 5 million square kilometres (1) whereas between 2006 and 2007 there was a sharp decrease (1) from 6 million square kilometres to 4.25 million square kilometres (1).
Or any other valid point.

(b) Increased temperatures are causing ice caps to melt so Polar habitats are beginning to disappear (1). Melting ice causes sea levels to rise (1) threatening coastal settlements (1). An increase in sea temperatures causes the water to expand, compounding the problem of flooding (1). Global warming could also affect weather patterns, leading to more droughts (1) crop failures and problems with food supply (1); flooding, causing the extinction of species (1) and more extreme weather, eg tropical storms (1). Tourism problems will increase as there will be less snow in some mountain resorts (1). Global warming could threaten the development of developing countries as restrictions on fossil fuel use may be imposed to slow the rate of increasing CO2 levels (1). In the UK, tropical diseases like malaria may spread as temperatures rise (1). Plants growth will be affected and some species will thrive in previously unsuitable areas (1). Higher temperatures may cause water shortages (1).
Or any other valid point.

10. (a) Overall the amount of deforestation in Peru 2004-2012 has decreased (1) from just under 3 million ha to 750 000 ha (1). The deforestation rate declined rapidly from 2004 to 2007 (1). Deforestation increased from 2007 to 2008 peaking in Peru at 1 500 000 hectares per year (1). Again Peru experienced a decline in deforestation rates from 2008 to 2009 by over 500 000 ha (1). From 2009 to 2010 deforestation rates rose to around 1 400 000 ha (1), before declining to around 750 000 hectares per year in 2012 (1).

(b) New industries have led to the expansion of towns such as Anchorage in Alaska which have grown to accommodate workers (1). Although these industries provide employment (1), these developments spoil the appearance of the natural landscape (1). New roads have been built to transport people and goods. This increases the number of vehicles in the tundra creating noise and air pollution (1). But also improves access to locals (1).

Oil is a very important industry in Alaska. The building of oil platforms and oil pipelines has resulted in damage to tundra vegetation and wildlife (1). In some areas, the Trans-Alaskan oil pipeline has been built on natural migration or hunting routes for animals, which hinders the natural movement of caribou (1). Local Inuit people have also had their way of life disrupted as they must detour around the pipeline (1) and may no longer have access to their traditional hunting grounds (1).

Local people were promised jobs in the industry, but few jobs are available for locals (1).

Burst pipes have spilt hundreds of thousands of gallons of crude oil in Alaska, devastating this fragile environment (1). Oil spills have also been responsible for pollution in the region (1), such as the Exxon Valdez disaster (1).

Any damage to the tundra landscape is slow to recover, as the short growing season means that bulldozer tracks from the oil and natural gas industries could take centuries to restore (1).

Pollution from mining and oil drilling has contaminated the air, lakes and rivers (1).
Any other valid point.

11. (a) Most cities are located on or near plate boundaries (1) where seismic activity is highest (1). Most earthquake threatened cities are found in developing countries (1) like Indonesia (1). A large number of threatened cities are found in China (1). Three cities in Africa are at risk (1). All threatened cities in the USA are found on the west coast (1) with a cluster around San Francisco/Los Angeles (1).
Or any other valid point.

(b) In Japan people take part in earthquake drills to practise what to do in the event of an earthquake (1) giving them a better chance of survival (1). The government warn people, using text messages and TV, giving them the chance to move to a safer place (1). Earthquake resistant buildings reduce the number of people trapped or killed (1). Sprinkler systems and gas cut off valves prevent fires spreading reducing the number of people injured and buildings destroyed (1). People living in earthquake prone areas have emergency plans in place and emergency supplies such as bottled water and tinned food are stockpiled to ensure they have vital supplies to survive in the event of an earthquake (2). In the event of an earthquake short term aid in the form of food, medicine and shelter is sent to the area to treat the injured (1).
Or any other valid point.

12. (a) The value of exports from developed world countries to developing world countries is $738bn (1) whereas there is only $650bn worth of goods exported from developing to developed world countries (1). That is a difference of $88 billion (1). The value of trade between developing world countries is $383bn (1). The value of trade between developed world countries is $2251bn (1). There is more trade between developed world countries than between developing countries (1); it is $1868bn more (1).
Or any other valid point.

(b) There is a big imbalance in the pattern of trade between the developing and developed world; this can reinforce differences in wealth between areas such as the EU and Africa (1); African countries export mainly primary products such as oil or cocoa beans for comparatively low prices but import mainly processed goods such as vehicles for much higher prices (1) which can result in a trade deficit for them (1); this can increase levels of poverty within African countries and cause difficulties for the economy as well (2); often the producers such as cocoa farmers in Africa receive very low wages and so struggle to maintain a decent standard of living (2); wealthy European countries profit from selling expensive manufactured goods to African countries (1), helping to keep a much higher standard of living for their citizens (1); often, exploitation of primary products in African countries can lead to serious environmental damage, such as logging which has caused deforestation (1), resulting in the loss of areas of rainforest as well as the destruction of animal habitats (1).
Or any other valid point.

13. (a) There has been a fairly steady increase in visitor numbers since 1995 (1) from around 525 million reaching 1 billion in 2013 (1). There were only 2 years where the numbers decreased slightly ie in 2003 (1) when it dropped to just under 700 million (1) and in 2009, dropped to under 900 million (1). The period with the largest increase was the 5 years between 1995 and 2000 (1) whereas the slowest increase has been in recent years from 2010 (1).

(b) If Costa Rica cloud forest chosen:
Eco-tourism raises local as well as international awareness of natural environment (1) such as wildlife and vegetation (1). Developing countries now want to conserve their fragile environments and view eco-tourism as a significant means of generating income (1). Developed countries want to help developing countries conserve their fragile environments by promoting sustainable/eco-tourism (1). Tourists are now more environmentally conscious and want to help protect fragile environments for future generations (1). Eco-tourism provides work and opportunities for local people (1) hence improving their standard of living (1) encourages local enterprise and improvement schemes (1) promoting awareness of local culture and traditions (1).
Or any other valid point.

14. (a) Male deaths from heart disease are most common in Eastern Europe (1). Russia for example, has a rate of 444–841 per 100 000 (1). This compares to only 120–238 in the UK (1). Canada, the USA and Mexico have some of the lowest rates (1), with under 120–238 per 100 000 (1). Many central African countries have rates of 363–443 (1).
Or any other valid point.

(b) If **pneumonia** chosen:
Antibiotics are used to treat any bacterial lung infections (1) and patients are encouraged to drink plenty in order to avoid dehydration (1); in severe cases a drip may be required to restore the right level of salts and fluids quickly (1); paracetamol is used to ease the effects of fever and/or headaches (1); introducing more community-based health workers helps to control the incidence of pneumonia as children with the disease are more likely to be diagnosed and treated quickly (1); this can often help to save lives (1). Vaccinations are being increasingly used in developing world countries to protect children against common infections such as flu (1); adequate nutrition helps to increase a child's natural defences against disease and so education about this also helps to reduce pneumonia (1).
Or any other valid point.

If **kwashiorkor** is chosen:
The main method of managing kwashiorkor is education about the need for a well-balanced diet, so that children don't develop the disease in the first place (1); by educating communities they can be encouraged to grow different food types to increase protein intake (1); this might include crops

such as cashews, peanuts, lentils or sunflower (1) and might also involve advice about constructing irrigation schemes to help crops grow better in times of drought (1); education about family planning also helps to reduce the number of children per family, making more food available per child (1).

For children who have kwashiorkor it is important to give vitamin and mineral supplements as salt and mineral levels in their blood stream may be dangerously low (1); Zinc supplements might also be administered to help the skin recover (1). Small amounts of food are reintroduced slowly, such as carbohydrates to give energy (1) and protein rich foods to help the child's body recover (1).
Or any other valid point.

If **malaria** chosen:
Anti-malarial drugs kill blood parasites (1). Chloroquine is an example of this (1). Insecticides, such as malathion destroy the female anopheles mosquito (1).

Draining all breeding areas eradicates larvae (1), planting eucalyptus trees to soak up moisture removes breeding ground (1). Water can also be released from dams to drown immature larvae (1). Mustard seeds can be used to drag larvae below the surface to drown them (1). Small fish can be introduced to eat larvae and provide a cheap protein source (1). Genetic engineering of sterile male mosquitoes reduces mosquitoes (1).

Health education teaches people about how to protect themselves from being bitten (1). Preventative bed nets are cheap and effective at stopping mosquitos biting at night (1). New treatments have also been developed which seem to be more effective such as artemesinin/ACT because malaria parasite is not yet resistant to them (1).
Or any other valid point.

If **cholera** chosen:
One of the main ways to reduce or control the spread of cholera is to improve sanitation which stops disease from spreading (1). Providing wells and pipes makes drinking water safe and clean (1). Health Education encourages people to wash hands often with soap and safe water preventing infection as does building and use of latrines (2). Because of contaminated water people should cook their food well and eat it hot (1). Food stuffs should be kept covered and fruit and vegetables should be peeled to prevent contamination (2).

Cholera is an easily treatable disease.
The main ways to treat cholera are either a simple drink made from 1 litre of safe water, 6-8 teaspoons of sugar and 1/2 teaspoon of salt, which helps to rehydrate sufferers so that they can fight off the disease (2) or re-hydration tablets, if available (1). In especially severe cases, intravenous administration of fluids may be required to save the patient's life (1). Treatment with antibiotics is recommended for severely ill patients to help fight the infection (1).
Or any other valid point.

NATIONAL 5 GEOGRAPHY 2016

Section 1: Physical Environments

1. (a) Pyramidal peak - 012216
Corrie - 800217
U-shaped valley - 927226

(b) A glacier forms in a corrie/north facing slope and moves downhill due to gravity (1), eroding the sides and bottom of the valley (1) through plucking and abrasion (1). This action makes the valley sides steeper and the valley deeper (1). When the glacier retreats a deep, steep, flat floored U-shaped valley is left behind (1). The original river in the valley now seems too small for the wider valley and is known as a misfit stream (1).

Any other valid point.

2. (a) Caves - 837160
Swallow hole - 891161
Intermittent drainage - 966146

(b) Limestone contains both joints and bedding planes, splitting the rock into well-defined blocks and making it permeable (1).

Water flows underground through a swallow hole, along bedding planes and down joints until it reaches impermeable rock (1). As it does so its slight acidity dissolves the limestone with which it comes into contact (1). A cave/cavern forms where there are many joints and bedding planes close together so that large areas of rock in the same space dissolve quickly (1). This leaves a large space underground which is called a cave/cavern (1). Some cave systems may also have been influenced by changes in the level of the water table and in volumes of water passing through as climate has changed (1).

Any other valid point.

3. A - coniferous woodland
B - minor road
C - Afon Mellte (river)

4. **Forestry:** (e.g. 8916)

This area is very steep and would be unsuitable for most other land uses (1). Much of the land is above 400m and is too cold for crops to grow (1). Soils might be acidic and rainfall is likely to be high, but coniferous trees can grow in these conditions (1). Many of the slopes are too steep to use machinery (1).

Recreation and Tourism: (e.g.8115 or 0020)

Limestone produces dramatic scenery such as limestone pavements (1). Tourists are attracted to the area for walking (1). There is a monkey sanctuary (1)

Glaciation produces high steep mountains and deep valleys which create dramatic views encouraging sightseeing (1). Steep corrie sides provide opportunities for rock climbing (1).

Farming: (e.g. 8815)

The land here is above 400m so hill sheep farming would be possible here as the animals can manage on the steep slopes (1). Valley bottoms could be used for cattle farming as the climate is warmer (1).

Industry: (e.g. 8319)

Limestone areas are sometimes used for the extraction of limestone (1). Quarries could be built here as there is limestone and also an A class road (A4067) nearby for the material to be transported (1). Opencast working shows evidence of industry, grid reference 8211 (1)

Water Storage and Supply: (e.g. 8321)

Glaciated uplands contain lochs (1) These could be used to store water and to supply water to towns and cities (1). These areas are high up and tend to have high rainfall to feed the supply (1).

Renewable Energy: (e.g. 0019)

Glaciated areas are high and exposed making them suitable for the creation of wind power (1). Winds are more common, with higher wind speeds likely, making them very well suited to wind power (1). Fewer people in this sparsely populated area may be affected by wind farm pollution (1).

Any other valid point or grid reference.

(If only one land use mentioned, maximum of 4 marks.)

5. **If Chart X chosen:**

A warm front has just passed over London, which will bring drizzle as shown in Diagram Q5A(1)

Manchester is in the warm sector, so it is warmer and drier than Glasgow, as shown on Diagram Q5A(1)

As Manchester is between fronts, the cloud cover there is less than Glasgow or London (1)

Glasgow is experiencing a cold front, so there is heavy rain as shown on Diagram Q5A (1)

The isobars are farther apart in London than in Glasgow and this explains the lighter winds in London (1).

A warm front has just passed over London, explaining the 7 oktas of cloud cover. (1).

If Chart Y chosen:

London is experiencing drizzle as shown on Diagram Q5A and this is possible in the warm sector (1)

The isobars are close together in the north, explaining the higher winds shown on Diagram Q5A(1)

As the warm front is yet to reach Manchester, Manchester is cooler than London (1)

The warm front has just passed over London meaning that it would be cloudy, accounting for the 7 oktas of cloud cover (1)

Any other valid point.

Section 2: Human Environments

6. (a) **Area X** - 4055 is the Inner City as it has a grid iron street pattern (1) old churches (1) old transport routes such as the railway used for industry (1) evidence of industrial buildings/close to docks (1) main roads leading to the CBD (1) as would be expected in the Inner City; this area is on the edge of the CBD which is where you would expect to find an inner city area (1).

Area Y - 3752 is modern suburbs as it has green space/woodland (1) it has a modern street pattern (cul-de-sacs) (1) It is located at the edge of the city as would be expected of the suburbs (1) there are two schools nearby for children of families living in the area (1) there are no main roads, only B class and minor roads (1)

Any other valid point.

(b) A - Museum
B - Ayres Quay
C - Docks

7. Improved diets such as those which include a variety of nutrients and protein help people to live a longer healthier life (1) eg as in Japan, where the life expectancy is 86 years of age (1) Access to a regular supply of clean water helps to reduce disease and death rates (1) Better pensions and good care for the elderly means that people are given the means by which they can live longer (1) Good sanitation has improved people's health which means that death rates are lower (1) Good medical care has improved peoples chances of maintaining good health thus reducing death rates (1) Vaccinations have helped reduce worldwide infant mortality (1)

Any other valid point.

8. **Diversification:** When farmers use other ventures such as farm shops it helps to boost the farmers' income (1) allows farmer to become more independent and less reliant on subsidies.(1) Visiting a farm means people experience rural landscape and outdoor activities (1) the farmer makes an income from accommodation, farm shops, farm attractions, tours, agricultural exhibits, wildlife tours, and country sports (1) Wind farm development on farming land also generates extra income (1)

Government Policy: In the UK, the Department for Food & Rural Affairs (DeFRA) or the Scottish Rural Development Programme (SRDP) supports farming industry by providing subsidies (1) DeFRA regulates policies which improve animal health and welfare regulations (1) Government demands disease control in plants and animals to maintain high standards of produce (1) Government funds and supports research into agriculture which in turn improves farming practices(1) CAP Common Agricultural Policy helps farmers to maintain stable prices and guarantee a steady income (1) farmers use set-aside land to prevent over-production of certain crops (1) Grants available for environmental improvements (1) such as planting hedges on rural land (1).

GM Crops: Genetically modified crops can increase crop yields (1) and improve resistance to disease (1) Many people disagree with GM crops arguing that it may have a negative impact on the natural environment (1). More tolerant crop varieties could be grown in areas where they couldn't be previously grown (1). GM crops reduce the need for pesticides which helps insects and bees (1)

New Technology: eg Using GPS to manage field operations or animal feeding saves time (1) computerised water management/irrigation can increase crop production (1) poly tunnels with environmental control systems can improve crop yield and quality (1) However cost of buying and maintaining this equipment and machinery is expensive (1) Chemical fertilisers and insecticides are widely used to improve production on farms(1) less labour required which has led to a decrease in population in rural areas (1). Overuse of chemicals may result in environmental damage (1). Drones may be used to survey fields of crops which helps farmers to quickly identify problems (1)

Any other valid point.

Section 3: Global Issues

9. (a) The northern hemisphere has experienced the most change in temperature (1) with some parts experiencing an average difference of 2°C (1). Parts of Arctic Canada have increased in temperature by an average of 2°C. (1). Most countries have experienced temperature increases eg USA, Australia and the

UK (1). The USA's temperature has increased by approximately 1°C (1). Brazil's average temperature has increased by 2°C (1). Some parts of the world haven't experienced a change in temperature e.g. Cape Horn in South America (1). Whereas, other places have experienced an overall decrease in temperature eg parts of Antarctica and the Southern Ocean: up to -2°C (2).

Any other valid point.

(b) Scientists observe and measure changes in temperature, CO_2 emissions and rising sea levels to monitor the rate of climate change and advise world leaders (2). Developed countries switch from fossil fuels to alternative sources of energy in order to reduce the amount of CO_2 in the atmosphere (1). Countries find new types of energy eg biofuels (1). Industries develop and expand existing sources that are more sustainable than fossil fuels eg solar, wind and wave power (2). Developing countries reduce deforestation and increase afforestation (1).

World summits enable governments to get together and discuss global strategies to try to reduce their use and consumption of carbon-based fossil fuels (1). Many governments signed the Kyoto Protocol, committing them to reducing greenhouse gas emissions (1). The UN climate summit in Paris December 2015 enabled world leaders to agree actions intended to avert the worst effects of climate change (1). Governments ban the use of harmful substances eg CFCs (1). The Carbon Credits Scheme is aimed at reducing greenhouse gas emissions by making the polluter pay according to how much pollution they generate (2).

London Congestion Charge: drivers pay for driving in the Congestion Charge Zone to cut the pollution generated from exhaust fumes (1). Industries and domestic users of energy are encouraged to use it more efficiently through media awareness campaigns (1). People are encouraged to walk, cycle, or use public transport rather than fossil-fuel powered cars (1). Bus lanes and cycle lanes designated to encourage people not to use their car (1). People use smaller more energy-efficient cars or electrical cars, helping to reduce fossil fuel emissions (1). Government tax is significantly reduced on vehicles with low CO_2 emissions (1).

Encourage people to holiday at home to reduce the number of aircraft journeys taken (especially short-haul flights) (1).

Educate people to switch off lights, power sockets, phone chargers and TVs when not in use (1). Recycle and reuse plastics and oil-based products (1). The Government now levy a charge of 5p for every carrier bag (1). Local councils supply bins to help householders recycle various products (1). Use energy-efficient light-bulbs and rechargeable batteries to conserve energy (1). Government grants to help home owners insulate house roofs and use more efficient heating systems (1). Install solar panels on house roof to generate renewable energy (1) or switch to an electricity supplier that supplies green electricity (1).

Any other valid point.

10. (a) **Answers may include:**
High rates of deforestation occur in Brazil, DR Congo and Indonesia (1). High rates of loss are also prevalent in areas such as Mexico and most of South America (1). High levels of loss are more common in developing countries (1). Moderate levels are common throughout Europe, northern Africa and Canada (1). Low rates are common throughout the USA, China, India and Australia (1).

Any other valid point.

(b) To ensure the tundra is conserved for future generations, sustainable development is absolutely crucial for its survival.

Management strategies include:

Habitat Conservation Programmes are sometimes established in tundra environments to protect the unique home for tundra wildlife (1).

In Canada and Russia, many tundra areas are protected through a national Biodiversity Action Plan (BAP) (1). The BAP is an internationally recognised programme designed to protect and restore threatened species and habitats (1).

Reducing global warming is crucial to protecting the tundra environment because the heating up of Arctic areas is threatening the existence of the environment (1). Most governments have promised to reduce greenhouse gases by signing up to the Kyoto Protocol (1).

Many countries have invested heavily in alternative sources of energy such as wind, wave and solar power. These sources of energy are renewable and more environmentally friendly than burning fossil fuels, which increase carbon emissions and global warming (2).

Some oil companies now schedule construction projects for the winter season to reduce environmental impact (1). Projects work from ice roads, which are built after the ground is frozen and snow covered. This limits damage to sensitive tundra (1).

Some oil companies locate polar bear dens using infrared scanners and do not work within 1.6 kilometres of these dens (1).

There are a number of Arctic research programmes, such as the International Association of Oil & Gas Producers' joint industry programme on Arctic oil spill response technology (1). This programme attempts to increase the effectiveness of dispersants in Arctic waters, oil spill modeling in ice and the use of remote sensors above and under water (2).

Many companies operate sophisticated systems to detect leaks (1).

Many companies work with local communities to understand and manage the potential local impacts of their work (1).

Many countries have set up national parks such as the Arctic National Wildlife refuge in Alaska to protect endangered animals in the tundra (1).

The Trans-Alaskan pipeline is raised up on stilts to allow Caribou to migrate underneath (1).

Any other valid point.

11. (a) Over the last 100 years the number of eruptions has increased from forty three in 1910 to seventy eruptions in 2010 (1). Apart from the decades of the 1920's, 1970's and 1990's the amount of volcanic activity in each decade increased (1). The least number of eruptions were in the 1920's with only

31 eruptions (1). There was a big drop between the 1910 decade and the 1920 decade with a drop of 12 eruptions (1). Also in the 1990's there were 12 fewer eruptions than the 1980's (1). The biggest increase was between the 1990's and the 2000's with 13 more eruptions (1). The greatest number of eruptions was in the 1980's, 2000's and 2010 at 66, 67 and 70 (1).

Any other valid point.

(b) For Pico de Fogo volcano answers could include:

The heat from the lava flows set fire to the main settlements destroying two villages as well as a forest reserve (2) endangering the vegetation and animal habitat (1). Around 1,500 people were forced to abandon their homes before the lava flow reached the villages of Portela and Bangeira on Fogo island (1). More than 1,000 people were evacuated from the Cha das Caldeiras region at the foot of the volcano to ensure their safety and prevent injuries (1). The airport was closed, as ash filled the sky, to prevent the risk of planes crashing (1). Buildings and records were destroyed resulting in some of the history of the area being lost (1). Roads and transport routes were destroyed affecting the tourist industry on the island (1). The volcano destroyed the agricultural land which resulted in the loss of fertile land (1) decreasing the ability of the area to produce crops (1) and support the local population (1). Tourism might increase as the volcano becomes a tourist attraction improving the economy of the island (1).

Any other valid point.

12. (a) Between 2001 and 2013 the export of goods to the EU has in general declined from around 60% to 50% (1). The number of exports to the rest of the world has gradually increased from 40% to 50% (1). By 2013 exports are now equal at 50% (1). In 2005 there was a drop in exports to the EU and an increase in exports to the Rest of the world by 2% (1). In 2006 there was a large increase in exports to the EU reaching its highest at 63% and a corresponding decrease in exports to the Rest of the World reaching its lowest at 37% (2) this being the biggest increase/decrease of 5% over the period (1).

Any other valid point.

(b) If the EU chosen:

Advantages:
The EU allows free trade within member states which allows all companies to trade on an equal basis (1) The EU creates more trade within its member countries (1). Consumers have lower prices, more choice and opportunities for work throughout the EU (1). Businesses have more consumers and are able to exploit economies of scale (1). The single currency, the Euro, means that it is easy for consumers to compare the price of products so makes markets more competitive (1). Poorer areas in a country can receive grants to improve the area (1) Free movement of labour allows people to work in any other member country (1).

Disadvantages:
Countries have to follow EU decisions/policies eg Common Agricultural Policy and decisions/policies made may not benefit all countries (1). High unemployment and low wages in new member states can lead to increased immigration but receiving countries may need to support them financially putting a strain on their economies (1). Increased tension can occur between immigrants and locals over jobs, housing etc (1). Countries have to contribute a set amount of money each year to a central fund(1).

Any other valid point.

13. (a) China's tourism expenditure has increased from about $128 billion in 2013 to $158 billion in 2016 (1) which is a considerable increase of $30billion in 3 years (1). Tourism expenditure in the USA increased by $2billion from $86 - $88 billion (1). Germany's tourism expenditure however decreased by $1billion from $86 billion (1). Tourism expenditure in Russia has increased by $11billion from $53 - $64 billion (1). Tourism expenditure in UK increased from $52 - $54 billion (1). General trend shows that tourism expenditure increased between 2013 - 2016 in almost all countries shown on the table (1).

Any other valid point.

(b) If the Caribbean is chosen:

Tourist industry aims to use social and environmental practices which benefit communities by protecting their environment and their heritage(1) eg Nature Conservancy Caribbean Challenge is an initiative set up to protect the Caribbean (1). So far 50 new marine/coastal protected areas are designated (1). The aim is to conserve at least 20% of their marine and coastal environments in national marine protected areas by 2020(1) aim to get the 40 million tourists who visit the Caribbean to help donate to the cause(1). A project in Jamaica aims to clean, upgrade and maintain resort towns(1) to increase security presence in order to reduce visitor harassment (1). In Dominica, the aim is for tourism to have as little harmful impact as possible on unspoiled areas of natural beauty (1). Solar power is used and water supply is pumped from the river using a silent solar powered pump, to avoid disturbing the surrounding natural habitats (1). To minimise water consumption, grey water is treated and then re-used in the garden and campers use dry toilets (1). All kitchen and garden waste is used as compost to grow as much organic food as possible without the use of chemicals or fertilizers (1). Whenever possible, the hotels avoid purchasing packaged goods and shop locally (1). Hotels recycle and use biodegradable products, and try to keep waste products to a minimum (1).

Any other valid point.

14. (a) Answers may include:

In April 2014 there were few cases of Ebola in Africa. By October 2014 there were almost 2500 cases in Liberia (1). In Sierra Leone there were almost 1,200 cases by October 2014 (1). In Guinea there were around 800 cases by October 2014 (1). In Liberia cases rose rapidly from around 250 in August 2014 to around 2500 by October 2014 (1). Sierra Leone witnessed a rapid increase in cases from around 500 cases on October 1st 2014 to almost 1200 by mid October 2014 (1).

Any other valid point.

(b) Answers may include:

Heart Disease:

Lifestyle factors are the main cause of heart disease. Many people do not take enough physical exercise which is necessary to keep the heart healthy (1). In developed societies many people take the car or use the lift rather than walking/taking the stairs (1). Poor diet also leads to heart disease (1). Too much saturated fat can cause hardening or blocking of the arteries (1). Many people do not eat enough fruit or vegetables, this can contribute to heart disease (1). Eating too much processed food, with a high salt content can also contribute to heart disease (1). Smoking can increase the risk of heart disease (1). High stress levels also contribute to heart disease (1). Possible effects of hereditary factors (1).

Asthma:
Infections such as colds or flu affect the lungs and narrow the airways, making asthma worse (1). Allergic reactions to dust mites in the home can cause asthma (1). Pollen from plants outside can cause asthma (1). Traffic fumes in polluted towns and cities can cause asthma (1). Cigarette smoke can also cause asthma (1). Asthma can be caused or made worse by damp conditions in the home (1). In cases of severe dampness, mould spores may help to make asthma worse (1).

Cancer:
An unhealthy lifestyle is the root cause of about a third of all cancers (1).

Smoking causes almost all lung cancer (1). Poor diet has been linked to bowel cancer, pancreatic cancer and oesophageal cancer (1).

Heavy drinking may be a factor in the development of cancer (1).

Some people may be genetically predisposed to some cancers, eg breast cancer (1).

Too much exposure to the sun can cause skin cancer (1). Obesity has also been linked with increased cancer risk (1).

Any other valid point.

Acknowledgements

Permission has been sought from all relevant copyright holders and Hodder Gibson is grateful for the use of the following:

Image © Alan Scott (http://www.RampantScotland.com) & Gianni Versace S.p.A. (2014 page 6);
Image © Ben Cooper (2014 page 6);
Image © Alis Leonte/Shutterstock.com (2014 page 13);
Image © Simon Rawles (2014 page 17);
Image © jan kranendonk/Shutterstock.com (2014 page 19);
Image © Gary Whitton/Shutterstock.com (2015 page 10);
Image © Kanokratnok/Shutterstock.com (2015 page 10);
Image © markobe/Fotolia (2015 page 11);
Image © Ivan Cholakov/Shutterstock.com (2015 page 12);
Image © Lledo/Shutterstock.com (2015 page 12);
Image © Phovoir/Shutterstock.com (2016 page 5);
Image © Sunderland University, taken from http://www.sunderland.ac.uk/images/Sunderland_City_Aerial.jpg (2016 page 9);
Image © Kostsov/Shutterstock.com (2016 page 13);
Image © LivetImages/Shutterstock.com (2016 page 17);
Ordnance Survey maps © Crown Copyright 2016. Ordnance Survey 100047450.